BODIES

BODIES

Why We Look the Way We Do
(And How We Feel About It)

BARRY GLASSNER

G. P. PUTNAM'S SONS / NEW YORK

Published by G. P. Putnam's Sons,
200 Madison Avenue, New York, NY 10016.
Published simultaneously in Canada

The text of this book is set in Caslon.

Library of Congress Cataloging-in-Publication Data

Glassner, Barry.
Bodies: why we look the way we do (and how we feel about it)/
Barry Glassner.—1st American ed.

p.   cm.
Bibliography: p.
Includes index.
ISBN 0–399–13342–9
1. Body, Human—Social aspects.   2. Physical fitness—United
States—Social aspects.   I. Title.
GT497.U6G56   1988   87–30524 CIP
391′.6—dc19

Printed in the United States of America
1   2   3   4   5   6   7   8   9   10

To Betsy

Contents

PART I

THE TYRANNY OF PERFECTION

ONE

Daily Lives

Take this morning, for instance. You turn on the TV at 7:10, and Willard Scott says it's going to rain again in the Midwest and there'll be a celebration of the American Eagle in a little town out in Wyoming. When he breaks for a commercial, the daily invasion of perfect bodies begins. Before the commercial starts, a handsome man and a beautiful woman appear on the screen. They're stars in a new hit TV series, and a voice in the background promises, "When we return . . . more on how they manage to look so terrific."

Then the commercial, which features soft music, perhaps Chopin, and a woman who is probably forty but has the snowy skin, exemplary figure, and long black hair of a twenty-year-old. She sits in an expensively appointed living room and expounds on how a new antacid fits into the new woman's life-style.

Switch to ABC. A feature story from California. A gorgeous blonde woman narrates as a dozen male models, eighteen to fifty years old, display new fashions for men. The narrator tells us that Robert Redford wears the first style, the second is popular at an exclusive club in Manhattan, and all the shirts are cut to show off well-developed chests.

Time for breakfast. On the cereal box, gazing right into your eyes, are a slender man and an even slimmer woman in their early twenties, sexy and smiling, wearing bright summer clothes. Open the newspaper and there's a Cher look-alike, only with larger breasts, sitting with her long legs outstretched and sporting a leotard cut to show most of her taut bottom and back. "You can have the body you want!" says the headline of this ad for a new exercise spa, which features "the European passive electronic technique." For just $59 a month you can "tone your muscles, remove stretchmarks, lose inches in 45 minutes in a workout equivalent to 1000 situps, 1000 pushups and 1000 leglifts."

The mail arrives with a brochure from a local night school offering courses like "Learn to Play the Guitar" and "Get Rich with Real Estate." The cover, which has nothing to do with the courses, shows a smiling, shapely young woman in a wet, low-cut bathing suit, posed in an old school desk chair and glancing at a book. "Summer School Was Never Like This" reads the headline beside her left hip.

The landscape has been overrun by pictures of attractive people. During an ordinary week most of us see several thousand exemplary bodies on billboards, in magazines, on television. Even before we finish our corn flakes we've compared ourselves, our friends, and our spouses to several of them, consciously and unconsciously; and we always come up short. The absurd disparity between viewer and image is suggested by a survey of the readers of *Vogue* magazine. Forty-nine percent wear size 12 or larger.

Why, at the end of the twentieth century, is the perfect body so much a part of our everyday lives? Why do we turn green with envy (and pink with guilt) when someone we know starts to look more like those images than we do?

To answer these questions it's essential to look beyond the standard explanations. I'm thinking of two in particular. The first, from those who worship the image, is that the current lean, exercised look is objectively a good, healthy standard. The second, from those who put themselves above it all, is that the ideal body of the 1980s is just a passing fancy, built upon fads.

I've often heard that latter view espoused by those who feel defensive about their own inability to participate. But to write off something as a fad is to beg the interesting questions. Why is it so popular? What does it tell us about our society and ourselves?

I'm going to argue throughout this book that images of perfect bodies captivate us precisely because they are *not* merely fads. Quite the contrary, they embody time-honored values of American culture such as self-control, personal achievement, and prosperity. As I will show, the ideal images have also become key parts of our economic system. They serve to channel our desires into the marketplace, where we spend an estimated $50 billion annually on diets, cosmetics, plastic surgery, health clubs, and gadgets.

Increasingly, the ideal images are also the standards against which our employers judge us. Corporations across the country have installed multimillion-dollar fitness centers on their premises and conduct weight-loss clinics during the lunch hour. Thus the ability to appear as lean, attractive, and youthful as the people in the ads becomes crucial for economic survival.

There is a great irony here. Even though the image is largely a fraud—people who pose for these pictures may neither look nor behave so enviably in real life—it's sufficiently powerful that it cannot be dismissed.

But should we go so far as to believe that a strong, fat-free body is obviously the best to have? I think not. The current ideal is not an objective representation of beauty and durability. Its assumed superiority is based on myths that have trickled down from cosmetics, exercise, and fashion moguls. We think it looks great because we've been taught that by so many authorities, from Mom to the American Heart Association.

If we're to understand why we are so worshipful of the lean, the young, and the well-exercised, we have to look deeper than conventional wisdom allows. Fortunately, important discoveries have been made recently by sociologists and psychologists who study the body ideals of our culture. We can now explain, for example, why people who are wealthier tend to be thinner, why men and women have different hang-ups about their appearance, and why few of those who begin an exercise or weight-loss program stay with it for more than a few months.

Body Talk

Concerns with staying thin and looking attractive have been documented in kindergarten children and in eighty-year-old retirees . . . and in every age group in between. To be sure, we're more focused on our bodies at certain times in our lives than at others, and the object of our attention often shifts. How, then, is it possible to get a grip on such a broad phenomenon?

The usual reply sociologists give to that question is: by means of surveys. And indeed, wherever I make a generalization about a group of people in this book, I have based it on either surveys or experiments. I have reviewed something like a thousand articles and books on the topics I'll be discussing and have conducted some surveys myself, with the help of research assistants.

But to rely too much on statistics in a discussion of this sort would be to contribute to the very problem this book is meant to address. Just as the ideal body in our society is not a real one but a homogenized and technologically produced image, so too are statistical profiles of the population. If all you want to know is the percentage of voters who will push the Republican lever in the next election, a large-scale survey of randomly selected citizens is fine, but if you hope to understand the lives and lusts of the lever pushers, you must talk with them directly.

That's why most of this book comes from confidential, in-depth interviews I conducted with ninety people throughout the country. They talked with me about how they cope, within the contexts of their relationships and their jobs, with the constant demands our society places on them to alter their appearance. Some, such as the actors, physicians, image consultants, and directors of health spas, are recognized as "experts" on the contemporary body by virtue of their technical competence. The majority of my interviewees, though, are experts in another sense: they are authorities about their own feelings and experiences.

I know none of the interviewees personally. I sought out a diverse group whom I located by means of assorted referral networks. I called upon current and former students, business associates, and family throughout the country, who in turn hooked me up with people they knew. This way each person I interviewed had some assurance, from a trusted acquaintance, that I could be counted on to protect confidentiality.

I have assigned pseudonyms to all of the interviewees and changed some telling information about their backgrounds or places of residence. In a few cases, I've merged two similar biographies into one. These alterations do not affect the points being made, but they do serve to protect the interviewees from being identified by employers or friends.

I wanted to get to know every person as intimately as possible. So in most cases I met with each one more than once, and always for two to five hours per visit. In about half the cases, I succeeded in interviewing both the person originally selected and his or her spouse or close friend. That way I was able to learn more about their social context, and to compare and check reports.

The interviews took me into a great variety of settings and lives: from a woman struggling with her declining beauty to another happy with her rotund figure; from men who pump iron to others who can barely lift their own luggage; from staid suburbanites to urban trendsetters; blacks and whites; service workers as well as executives and professionals.

In a modest home in the Wallingford neighborhood of Seattle, for example, I interviewed a pair of high school teachers: Carrie, thirty-two, and Bob, her forty-six-year-old husband, both of whom devote their lives to their work and to staying fit. They get out of bed at 5:15 A.M., do a few stretching exercises, shower, eat what Carrie described as "a milk-shake kind of thing with brewer's yeast and juices and vitamins and things like that," and drive forty-five minutes to be at work by 7:00. They teach until 3:30, when they drive to an exercise emporium, work out for two and a half hours and then drive home for spinach salad or stir-fried vegetables with multigrain bread and some vitamin pills. After dinner they grade their students' papers or read the new issue of *Prevention* magazine until 9:30 or 10:00, and then go to bed. They admire each other's body, and make passionate love several times a week (mostly on Saturdays and Sundays).

In contrast, but no less under the influence of contemporary expectations about the body, were Madeleine and Sean. Both in their mid-forties, and living in the LA suburb of Sherman

Oaks, they love each other emotionally, but physically they're worlds apart. Sean, a cardiologist, goes to bed early in order to rise before dawn for his daily five-mile run. Madeleine, a talent agent, stays up half the night reading scripts, and mounts her stationary bike, which she detests, only under pressure from Sean. He's lean, she's plump. He wishes she'd lose weight (so does she), she wishes he'd gain some (he disagrees). Both wish their sex life would improve.

Carrie, Bob, Madeleine, and Sean, like the majority of those I interviewed, are struggling, in diverse and fascinating ways, to meet society's expectations of how their bodies should look and be cared for. Others I interviewed are actively questioning society's models of perfection.

Take, for example, Sharon, a forty-one-year-old feminist English professor at the University of Texas well-known for her critical essays about how the beautiful women in novels and films have brainwashed the public. She spent her childhood battling her parents over her weight. Somewhere in late adolescence she started to rebel, and she's been doing so ever since.

"I'm angry at the images of attractiveness," she told me, "but nonetheless I do feel coerced and compelled by them. The best I've been able to do is to counteract that force with the side of me that resists being contained or suppressed.

"It's really a very painful conflict, because I'm constantly going back and forth and no state is satisfying. I absolutely refuse to go on diets, for example. But then at a certain point I get dissatisfied with myself. It's not as though the state of just eating whenever I want and doing whatever I want satisfies me. It doesn't. Because I'm constantly haunted by the fact that I am not as attractive as I want to be. The social standards are still my standards.

"The choices are rotten ones, and no one is happy with them. I see most of my undergraduate students choosing to be practically anorexic and trying to fit what they see in the magazines. It's the same for my women friends. In the sixties and seventies they tried the 'fat is a feminist issue' ideology, but in the eighties they got swept up by the 'fitness is a feminist act' mentality and now they're lifting weights. They want to look like models for fitness studios . . . just like everybody else in America."

A Long, Strange Trip

One of the more surprising moments in my journey occurred when I walked into Stuart's plush office in Atlanta and caught him scrutinizing the bathing-suit issue of *Sports Illustrated*. He didn't notice me, so engrossed was he by the photo of a big-breasted blonde wearing a tiny white bikini and stretched out seductively on a bed of rose petals. "I'll show those sons of bitches," I heard him grumble to himself.

Stuart is a prominent plastic surgeon. Fearful that he might be embarrassed or angry that I'd caught him in the act, I quickly approached his desk and began to introduce myself. "Look at this," he interrupted, running his index finger over the model's tanned posterior. "What a great shot. You can see within half a centimeter of her anus."

A little flushed, I took out my tape recorder and notepad and said how pleased I was to have a chance to interview him.

He held up the magazine for me to see the blonde's left bun under his index finger. "Now you find me somewhere I can take a skin graft without it showing," he said. "With today's bathing suits, it's impossible. I can't wait to bring this thing into court."

I said I didn't understand. He pulled his attention away from the magazine and explained that he was testifying the following day in a lawsuit brought by a wealthy woman who was suing a medical school chum of his for $2 million. She was upset that this plastic surgeon, who had reconstructed her left breast after a mastectomy, had left a scar on her buttocks that prevented her from wearing a bathing suit in public. Her lawyers were arguing that the surgeon should have taken the skin without noticeable evidence that anything had been done. As one of America's leading plastic surgeons, Stuart was an ideal choice to serve as expert witness on behalf of the doctor. In the course of our time together, he educated me about the ethics and aesthetics of sculpting people to look as beautiful or youthful as their lovers or bosses want them to look.

Then there was the interview I nearly got stripped for. My appointment was with Edward, the superintendent of a large prison in Maryland. At the metal detector near the entrance, I took out my keys and change. Still, the alarm sounded when I walked through. The guard asked for my suit coat and sent me through again. No luck. "Your belt," he said, the siren blaring.

Still I didn't pass, and the guard, annoyingly deadpan, called for my shoes and everything else from my pockets. I removed a small tin of aspirin and marched through the machine, confident this trip would be the last. The buzzer seemed louder. As the guard asked for my shirt and pants, I noticed just how open this entry area was. Luckily, when I unbuttoned the cuffs of my shirt the guard said coolly, "Your watch, sir," and that was that.

Fortunately, the interview with Edward was well worth the embarrassment I suffered on the way into his office.

Though his suit came from Sears and his hair was cut by a barber rather than a stylist, Edward led me to understand the nature of male vanity, and why it is usually kept private.

On the whole, men had a harder time answering my questions than women did, but in the end I was surprised to learn how ill-equipped we all are to talk about our bodies. English contains precious few adjectives to describe an attractive person, and how many synonyms can you think of for *fit*? Negative body types don't fare much better, with the important exception of *fat*, which can be expressed with many nuances (*chubby, flabby, pudgy, plump, stout, tubby*, etc.). Like the Eskimo who have twenty words for snow, we make fine distinctions about weight, the feature of our environment that concerns us most.

Even those who talked freely about the details of their sex lives blushed at some of my questions about their bodies. A case in point: Rosie, a twenty-six-year-old rock guitarist in New York, rambled on for hours about her affairs and each of the sexual positions she and her various male and female partners had improvised. Yet when I asked how she felt about her looks at various times in her life, or why she had gained and lost weight so many times, I had to press for a response.

The ideal images of bodies are largely to blame for our impoverished language. If there are only two categories of bodies—namely "attractive" and "needs work"—as the ads in men's and women's magazines imply, relatively few words are required to describe the fine points about either.

Despite some reticence, the conversations I had were deeply personal and revealing. At the end of each meeting I asked if I had neglected to discuss anything the interviewee considered important. Often the response was like Lynn's.

After an emotional four-hour review of her life, Lynn, a secretary in Indianapolis, replied to my question: "I can't imagine what you could have missed. After all this talking, you probably know as much about me as I know about myself."

Charlie said much the same. A handsome black man, he taught me how someone's life can be ruled by other people's stereotypes about his appearance. As I was about to leave, Charlie said, "It's funny. This is the first time I've ever involved myself in any kind of self-disclosure to this extent, but I really enjoyed it."

In the pages that follow, you will hear these people's stories—stories of their struggles to come to terms with the various expectations about their bodies, imposed on them by the culture, by their families and friends, and by their own self-images.

TWO

The Power of the Image

My travels took me into homes that ranged from Toyota-sized studio apartments in New York to enormous mansions in the hills around Los Angeles. Each was decorated to reflect the taste and discretionary income of its inhabitants. But as different as these places were, scarcely one did not have on display some images of beautiful people.

In a tasteful condo in central Philadelphia, a plump and lively hospital administrator named Jacqueline was telling me about herself, when suddenly she stood up and walked me by the elbow from her living room to the kitchen.

"A year ago I cut this picture out of *Redbook*," she said, pointing to a clipping of a model attached to her refrigerator door with a magnet that looked like a slice of cherry pie. "This woman was what I was going to get to. I didn't in any way believe that this body at age thirty-nine was going to look like this woman in the ad, who is probably nineteen. But I thought it might give me a little incentive to lose weight."

Did it? "No. I'm going to replace it with this sign my sister-in-law gave me." She reached beneath a stack of dish towels in a drawer and pulled out a three-inch-square red magnet emblazoned with the message: DON'T PIG OUT.

We both laughed. Jacqueline returned the sign to the drawer, and we walked back into the living room and sat down on her plush new sofa. She looked a little dejected. "I'm pretty happy. I have a good job, a decent marriage, more money than I ever dreamed of when I was growing up. But with my body . . . I read the women's magazines, I watch TV. I see these women, particularly in the ads, and I find them very attractive, I like to look at them, and that has a tremendous appeal to me, to look like that. But the reality is, I weigh 175 pounds and my face is pretty, but I'm nothing to write home about."

I might have expected such concerns from a fashionable woman with a weight problem living in a natty neighborhood in the Northeast. But who'd have imagined that a lean, mean, working-class, white midwestern man would have set his sights on looking like a TV star? In Indianapolis I met him. By day Pete worked as a custodian emptying wastebaskets behind desks and tidying up reception areas at a big corporate office building. After work he took off not for beers with the guys but for the Nautilus machines at the health club.

By the time we met, at 7:30 at night, he'd eaten a salad, moussed his hair, and decked himself out in designer corduroys and a fitted dress shirt that showed off his muscular build. He was about to head out on a date with a secretary he had met the previous week at the club. "She's going to be a professional model," Pete said as he led me over to the white counter that divided the living room of his modern one-bedroom apartment from the kitchen. There, beside the current issues of *Penthouse* and *Gentleman's Quarterly*, was a photo of Pete's date in a sleeveless top and short skirt, looking every bit as busty and fat-free as a *Cosmopolitan* model.

In a life zone far removed from both Jacqueline's and Pete's, in a fourth-story walk-up in lower Manhattan where she rehearses, Rosie, twenty-six—wearing a punky outfit of black trousers, sneaker-style black boots, oversized black-rimmed glasses, and a very long silver earring in her left ear—was practicing a chord over and over again on an electric guitar when I walked in.

She, too, despite her apparent nonconformity, grazes on images of perfect bodies. The current issue of *Details* magazine lay open on the floor, amid a heap of concert posters and paper napkins scribbled with lyrics. When I asked about her childhood she replied: "I wasn't real athletic or anything, which is funny 'cause now I am. I go to the gym like five times a week. I guess I started 'cause of the stress of living in New York. The reason I keep going is that I want to have a good body. I want to be in shape. See, my mother's obese, and I really have this thing that I don't want to be fat.

"I think I would lose a lot of self-respect if I didn't look good. I know it's vain, but everyone wants to feel attractive. A large part of your identity comes from your image. I dress in a certain way to give a certain image, and I want my body to give a healthy image, too."

Almost every sector of American society now worships at the altar of the ideal body. Yet I can remember that not so long ago there were atheists and freethinkers who refused to kneel. During the sixties and most of the seventies, feminists shunned the cosmetics industry and urged the acceptance of all women as beautiful, whatever their jeans size. But by the beginning of the eighties, objections to conventional standards of beauty gave way to the idea that one can still be a good feminist and inspire construction workers to whistle. In the issue of *Ms.* magazine that arrived as I was writing this chapter, a majority of the display advertisements were about staying beautiful, fit, and sexy.

Nearly every model in these ads was thin, gorgeous, and young.

On the male side, I can remember the days when men were men and boys were hippies, when no self-respecting working-class man would openly preen and no member of the Woodstock generation would sport developed biceps or professionally styled hair.

But nowadays, not only have former bra-burners and counterculturalists rationalized their quest for a better body, but even the younger generation of rebels has done so. Said Rosie, "You run into a problem where, okay, there's this idea that you are like decorating yourself for the male and you shouldn't do that because you're a victim of a patriarchal culture. I think that's valid, but I also think that women have to reclaim their own health and self-esteem and decoration, because that's a normal human instinct."

Rosie can work on her Rosanna Arquette–like figure to allay her fears of becoming like her mother, yet still be 100% Downtown Girl. Of course, she would never do the yuppie activities at the gym: "Kick left, kick right, like a machine," she complained. Instead she runs laps and pumps free weights off by herself to New Wave tapes on her Walkman.

Likewise her preferences in men. Whereas she once pursued men whose physiques were as idiosyncratic as their politics, now she accepts the societal ideal as her own. "I like slim men with tight asses and fairly broad shoulders, taller than I am, and fairly good muscle tone," she told me. Lest she sound conventional, she quickly added: "I don't mean Wall Street types or anything. Suits turn me off. Most men I look at are wearing earrings and radical pants. But have you seen that guy on the Calvin Klein billboard? I think he's delicious. I just wish he wouldn't wear those awful clothes."

The Act of Looking

Why do these images of beautiful people hold so much appeal? A very basic answer—and a correct one—was offered back in 1935 by the psychoanalyst Otto Fenichel. "One looks at an object," he wrote, "in order to share in its experience." While we're consciously thinking about how gorgeous the person in the picture looks, *un*consciously we're identifying with him or her. The good feelings that come from looking derive not from some pure aesthetic experience, not because looking at something beautiful is naturally and simply pleasurable. We feel uplifted or sexually excited because we momentarily imagine ourselves as actually being that beautiful person, or having that person as a lover, complete with all the power and sensuality that particular body conveys.

And yet our response is not entirely positive. Although we enjoy and identify with the object of our gaze, we also hate it. "Very often," Fenichel wrote, "sadistic impulses enter into the instinctual aim of looking: one wishes *to destroy* something by means of looking at it, or else the act of looking itself has already acquired the significance of a modified form of destruction."

Evidence popped up frequently in my interviews of the subterranean anger people harbor toward ideal images from the media. Several told me of secret urges to deface billboards. Rosie described a long dream she'd had a few weeks before we met, a dream almost too hip to be true, in which she chased after Marilyn Monroe in a wheat field, finally assassinating her with a machine gun. For having committed this crime, she was hunted down by a hundred men and sentenced to spend the rest of her life in Bulgaria.

I suspect this pervasive love-hate relationship is a distinctively modern affliction, even though idealized images of the body have been present in the West since at least the fifth century B.C., when Greek artisans first produced sculptures of men, then of women. People were astounded at how lifelike these creations seemed, but in fact the statues were glorifications. Bodily imperfections were smoothed out, proportions improved, sexual parts made even and hairless.

Since then, images of Venuses and Adonises have been produced and displayed for centuries, but it is only relatively recently that people have become especially susceptible to their influence. Social historians point out that for most of human history, visual perception was much less important than it is today. As late as the 1700s people tended to trust what they heard or felt over what they saw. Touch was thought to provide more direct contact with something than sight could, and in an oral culture in which almost everyone was illiterate, knowledge about the world typically came by way of oral reports.

It was the printing press that first made the sense of sight so central. Unlike stories passed from one person to another and changed in the process, printed information remained the same from one printing to the next and therefore came to take on greater worth than that conveyed by touch or sound. As people grew more sight-conscious, they became more concerned with how they themselves looked, and popular treatises were written about how to dress and carry oneself.

A real turning point occurred around 1850, when the photographic negative came into use, and mass production of photographs began. In America, the middle and upper classes developed a passion for collecting photographs, to the extent that picture albums became, as some have put it, the television of the Victorian home. People bought pictures of performers, politicians, and circus curiosities; families spent hours arranging photographs and looking at them; and pictures of the powerful, beautiful, or famous began to serve as models for how to look and behave.

By the 1920s, pictures had altered the way in which people looked at the body. The body in a photograph is immediate, posed, frozen, unlike the live body with its constantly changing profile. Under the influence of photography, the instantaneous impression was born, exemplified in the changing fashions of the period. Styles shifted from clothing "made of many separate parts, each individually designed and each needing separate, slow appreciation," to "the new total look, instantly perceived," reports Anne Hollander in *Seeing Through Clothes*.

The increasing dominance of the sense of sight was also reflected in the rapid growth of the optical goods industry. Eyeglasses had been available since the sixteenth century, but not until the mid-1800s was significant progress made in testing and correcting eyesight. Then, between 1880 and 1929, the sale of eyeglasses increased roughly thirtyfold. "This meant," historian Hillel Schwartz points out, "that the visual blur with which millions of Americans had lived was no longer acceptable."

At the same time that Americans became more visually oriented, they also became more product-oriented—and the body took center stage in a new urban drama.

Historians have pointed to the department store—which came of age in the nation's cities in the late 1800s—as a milestone in the process of turning us into a country of consumers. Housed in sumptuous and showy buildings, department stores delivered a clear message to the middle class: the lifestyle of the rich and glamorous is available to you. You can elevate your station in the eyes of others by buying the right decorations and displaying them in your home and on your person.

Previously, people had sized up one another's social status and desirability on the basis of their family background. But in the growing American city, lineage no longer meant so much. Children and grandchildren of immigrants were becoming wealthy industrialists. The major clue to status became the image people conveyed.

"Individuals had now to decode . . . the appearance of others and take pains to manage the impressions they might give off, while moving through the world of strangers," says cultural historian Mike Featherstone. "This encouraged greater bodily self-consciousness and self-scrutiny in public life."

Over the course of the twentieth century, image has come to be as important as ability. As Christopher Lasch has said: "Both as a worker and as a consumer, the individual learns not merely to measure himself against others but to see himself through others' eyes. He learns that the self-image he projects counts for more than accumulated skills and experience." If that sounds like an exaggeration, just take a look in the marketplace. A help-wanted ad in *The New York Times*, for an administrative secretary who will be paid $40,000, clearly states: "Your good looks win here!" Another, for an administrative position in the $22,000 range, specifies, "clean-cut, 3-piece preppy look."

Product People

These days we've grown so image-oriented that we interpret one another by way of what might be called consumption cues. In an experiment at Arizona State University, for example, students were presented with the beginning of a short story and asked to complete it. Some were handed a version that described the protagonist as a man who used Crest toothpaste and drove a Honda Accord, while others were told that he brushed with Ultra Brite and drove a Camaro.

The story itself was about a first date. The young man

and a woman had gone to a movie, he'd driven her home, and they were parked behind her apartment. "The next move was his," the tale continued. "What to say, what to do? It may have been only a first date, but he was not willing to let it end. But there would be other dates with her. He was frozen in self-debate. After a few agonizing moments . . ." The students were asked to complete the sentence and the story.

The Crest–Accord man much more often was described as ending the date with a "Good night," a handshake, or a kiss. The Ultra Brite–Camaro fellow, in contrast, was expected to go for more than a kiss. These expectations were held by male and female students alike.

What clearer proof do we need that characteristics associated with a product are attributed to their users? At the time the experiment was conducted, the Camaro was being advertised as "an aggressive new road car" with "an unusual capacity for enhancing all but the most conservative of lifestyles"; and Ultra Brite was calling itself "the sex appeal toothpaste." In contrast, Honda described the Accord in terms of its "efficiency and commonsense advantages," and Crest was promoted as the toothpaste "for improving your family's dental health."

The economy itself has come to be based on such images. If everyone were satisfied to be either a Camaro or an Accord type of guy, or remained loyal to last year's styles in cars, clothes, cosmetics, and calisthenics, a crisis greater than the stock market crash of 1929 would ensue. It is fundamental to our economic system that we constantly want more and different things to rejuvenate our self-image.

Advertisers both respond to and help create the terms within which we think about our wants. A team of Canadian professors has catalogued the relationships between people and products implied in advertisements throughout this century. In the early years, the major appeal was practical: the objective qualities of the product and its utility for the buyer were the main focus. But beginning in the 1920s, "emphasis on what the product *did* diminished, while the visuals increasingly explored what the product could *mean* for consumers." Ads were about what kind of man or woman used the product, and the importance of the product in their lives. A 1925 Ford Motor Company ad in *Ladies' Home Journal*, for instance, featured fashionable women getting out of their $520 car beside a golf course. The text read: "Your Heritage of Health. Guard it with exercise in the open. A Ford car has been the stimulus to thousands of women to lead happier, healthier, more active lives."

The consumer was brought into an increasingly intimate relationship with the product during the decades that followed. The Canadian researchers found that the themes of sensual appeal and self-transformation appear more frequently in ads in recent decades and that ads have become more "personalized." People give testimonials about the product, or lend an image to it, as epitomized by the Marlboro Man, whose presence makes the cigarette (and by implication, those who purchase it) more macho. In other instances, the product is actually treated like a participant in the human world. In one Chevrolet campaign, the automobile in question was referred to as "more than a car—a member of the family."

Advertisers also encourage us to associate their products with fears and fantasies we harbor about ourselves, in particular about our *physical* selves. A message of many ads throughout the twentieth century has been: *You need* our *commodity for the sake of* your *most basic commodity: your body*. In the words of a Camay soap ad from the 1930s, "life is a beauty contest," and you can win it "if your skin is lovely and deserving."

Rare is the ad for cosmetics, clothes, or exercise clubs that does not harp on our physical insecurities.

Automobile ads, too, take a corporeal approach, hawking cars as add-on bodies. Ford tells would-be he-men they'll be "tough trail bosses" with the "powerful new V-6" and "strong suspensions" of the baldly named Bronco, a vehicle shown tearing through marshes. Even Honda, in trying to sell its Prelude to image-conscious executives, offers "muscle," "finesse," and "elegance."

Ditto for ads trained at women. Consider, for example, Ford's offer to the young woman who worries whether she can attract a man in the fiercely competitive mate market: Mercury is "the shape you want to be in."

Given the takeover of our lives by surface appearances, and the goal of advertisers to make us feel intimately connected to their products, it's no wonder that so many ads feature beautiful people. Sometimes these are imaginary people, as in a famous Ajax commercial of the sixties, with a strapping white knight on a charger. One of the creators of the ad was quoted as saying at the time, "Every housewife has been waiting for a white knight since she was a little girl. When we say 'stronger than dirt,' we are saying to her, 'stronger than your husband.'"

The imaginary person is a rarity, though, in contemporary advertisements. Instead, ads are populated primarily with the young, the attractive, and the lean. With ads costing as much as $13,000 per *second* on network television, advertisers require hard-hitting and memorable messages. Those who produce TV commercials and magazine layouts know instinctively what social psychologists have shown empirically: we respond more enthusiastically to attractive people than to ordinary people. Why this is so has been hotly debated, and researchers suggest several explanations. The most obvious possibility, "beauty commands attention," is probably false. The media are so full of beautiful people that to catch the viewer's eye one would do better to hire an ugly or unusual model (and a few advertisers have taken that approach, perhaps the most memorable being Wendy's "Where's the beef?" campaign). We remember both unattractive and highly attractive faces better than average faces. But then, the advertiser's goal is usually not simply to make us look and recall; it's to provoke a feeling of desire, and pretty bodies are capable of that.

According to surveys, attractive models in ads serve as our models in life. Most people assume that those who are endowed with beauty also possess the other good things in life. The highly attractive are presumed to be healthier, saner, happier, and wealthier than everyone else, as well as more confident, poised, trustworthy, and powerful. Place your product on, next to, or under a perfect body, and you just might be able to convince consumers that these qualities will rub off on them.

But how accurate are these stereotypes about the good-looking in the first place?

A variety of studies indicate they do have some basis in fact . . . but only *some*. A national survey discovered that the average income of attractive men and women is 11 percent higher than that of regular-looking people, and 19 percent greater than what plain or homely people make. Researchers have also found that attractive folks less often end up in psychiatric facilities, and that they consider themselves happier than do average or unattractive people.

On the other hand, although beautiful children often receive special attention from teachers and other children, less attractive kids typically score higher on IQ and college admission tests. Evidence also suggests that people who are good-looking often have difficulty sustaining friendships and love relationships. And when given tests to measure their self-esteem, they come out only about equal with other people.

For both men and women, the benefits of being beautiful are unreliable. Statistics show that average-looking men end up with wives with more years of education than do highly attractive men, as well as more years of education themselves and more prestigious jobs. Less attractive men also have an advantage when seeking friends. In an experiment in which men were shown photos of other men and asked whom they would like to meet, they more often selected not the best-looking, but those somewhat less handsome.

Attractiveness might seem less of a mixed blessing for women than for men, but studies suggest otherwise. Pretty girls do tend to be more "popular" while growing up, and they end up with wealthier husbands. On the other hand, men who seek out beautiful women are not necessarily those with the most attractive personalities. In one study, young men were asked to choose a date on the basis of pictures of women and accompanying information about their interests. Some selected primarily on the basis of how attractive the woman was, others on the basis of her interests. Those who picked prettiness tended to be "high self-monitors"—the type who want very much to control the images they project. In contrast, those who picked personality were "low self-monitors"—more aware of their own feelings and less interested in putting on an act for an audience. The attractive attracted the self-absorbed, while the rest attracted the self-assured.

More so than their male counterparts, beautiful women are stereotyped as oversexed. Men catalogue them as dangerous, and women convert jealousy into disapproval or assume that since beautiful women spend so much time being pretty, they can't be very accomplished in other ways. Erica Jong has pointed out that women were accused of witchcraft not only for being ugly but also for being gorgeous.

Surveys indicate that women who were especially attractive in college are less satisfied with their lives at middle age than women who were less attractive. My interviews suggest this is only partly because they feel they have lost something, since frequently these women are still considered highly attractive later in life. Rather, it seems that almost all the rewards of beauty—the dates or husbands, the special treatment from teachers and employers—have been received by the time a woman is thirty, and it has become a chore to look attractive all the time.

This holds true even for a woman of fifty-two who is pretty enough to make her living from her appearance, as I discovered when I interviewed Mimi, a successful model.

Selling Yourself Like a Product

In her late teens and early twenties Mimi had the sort of life other girls dreamed of: an apartment in New York, handsome men begging for her attention, a successful modeling career, plans to become an actress. Nothing dramatic intervened, no life-threatening disease or auto accident, just a gradual loss of interest on Mimi's part. At twenty-two, as she started to tire of the hundreds of auditions and poorly paid off-Broadway roles, and her modeling portfolio began to grow stale, one of the handsome men who courted her persuaded her to marry him. Luckily, he also happened to be rich.

To make a long story short, Mimi had three children, a lousy marriage, and some decent affairs. She married one of the men she saw on the sly as soon as she and her first husband split up, and she went on to have two more children with him. The youngest was in grade school when Mimi got divorced again.

She's still close to her five children but not to their fathers. Nor, for that matter, does she speak to her third ex-husband, with whom she broke up twelve years ago, at the same time she returned to modeling.

Sitting in her large Upper West Side apartment, in a living room filled with assorted paintings and theater posters and barbells she uses to keep her upper body tight, Mimi seemed a little annoyed when I asked whether she thought it was hard for the average woman her age to meet the beauty standards imposed upon her by the media. "I was born with good looks, but the reason I look younger than most fifty-year-old women is that they let themselves go to hell and I don't," she said. "Even still, I have a pot. I used to do two hundred sit-ups every morning to get rid of it, but I've had five babies and it's not going to go away. They can hide it pretty well when I'm photographed."

I asked Mimi what she does now to keep in shape, and she launched into a long testimonial about running. The more she talked about it, though, the more evident it became that she was running for reasons other than her appearance. Her fifty miles per week had actually *hurt* her modeling career. In each of the six years she'd been running, she had suffered at least one serious bone or nerve injury which had kept her out of work. And she frequently looked too gaunt to photograph well.

Mimi told me she resents it when something noble, like running, interferes with something trifling, like modeling. "I should be ten pounds lighter for quality running," she said. "I think I look terrific when I'm very thin. But casting directors don't agree."

Mimi runs, she intimated, because when she does she's no longer just an aging woman who supports herself with modeling while auditioning for television and stage roles she never gets.

"I win when I run," she informed me. "I win medals, I win plaques. I had never been a great athlete, but I'm a great runner. It's good for my ego. When I run in Central Park people see me and join me. Even people who run faster than I do slow down and run with me. I look pretty good those days, I guess, and people look. There are days when every man who goes by will smile, which I get a kick out of."

She wasn't saying running is *fun*, she clarified immediately. In Mimi's view, some things in life are really worthwhile (in particular, acting, raising children, and running) and therefore deserve hard work even if the payoff is limited, whereas others (like marriage and modeling) merit little effort, and you take what you can get from them.

She's gotten a lot from modeling—about half a million dollars between her fortieth and fiftieth birthdays. Recently, though, her career has taken a dip. Her marathoner body is one reason, but there's another as well. "I have gone through several years of doing less work than I would like because I look the wrong age," Mimi explained. "I'm too old to be a twenty-eight-year-old mommy of a four-year-old, and I look too young to be the senior citizen with arthritis. There's not much work for anyone in between."

People who are middle-aged—like people who are poor or handicapped—are largely invisible in advertising. They may make up part of the market for a product but are usually not part of the image the product is intended to convey. About two-thirds of the models in magazine ads are young adults, according to recent counts. Middle-aged women violate the myth that everyone can be young until becoming truly old, and that beauty is something available to anyone who will work for it. To display a model as old as Mimi as overtly middle-aged would be to remind the consumer that, no matter how much she spends on running shoes, cosmetics, clothing, or beauty programs, there's no cure for growing older.

The shortage of modeling work for models Mimi's age mirrors the invisibility imposed upon American women from their late forties to their sixties. They are expected either to deny their age and devote themselves to feigned youthfulness (become thin and sexy, the magazines tell them) or else to hide out in an office job. Or they can prepare for grandmotherhood early.

Mimi, however, doesn't concern herself with such sociological issues. When I asked how she felt about losing work because of her age, she replied, "You don't survive in this business if you take it personally. When I started back in modeling and I was still getting the instrument working and the juices flowing again, I was seeing a very smart CEO. This guy had worked his way up from traveling salesman. He had a motto: 'Sell yourself like a product, like you're selling shoes, and don't take it personally when they don't buy.' "

Mimi can live with rejection and less work. What really gets her goat is what photographers do to her when she *does* get work. She almost never portrays a woman of her age and background. In the ads she's always middle-American pretty, and either a mid-thirties businesswoman or happy housewife, or else the younger woman walking arm-in-arm with the older man.

She doesn't object to role-playing in itself; she's an actress, after all, and perfectly capable of putting aside her Italian roots to portray an overgrown WASP cheerleader from Omaha. It's the absurdity of how she is displayed in those roles that frustrates her.

"It always amazes me," she said, "that they want me because I look fit and healthy, and then they dump so much crap on my face I look like I'm ready to be buried. The other day, I did a spot for some kind of frozen diet dinners. They teased my hair and put on enough makeup for ten women."

Mimi's sentiment is one even younger models express. Monika Schnarre, who made it to the big leagues of modeling while still in junior high school, remarked at age fifteen to *People* magazine: "How ironic this all is. I'm hired for my looks, and yet it takes them three hours to make me pretty enough to photograph. Isn't that weird?"

Actually it's not, given the role that ideal images play in the culture. The idea is not to present the model herself but to make her into an assemblage of purchasable products, as the cover credit from a recent women's magazine clearly states: "Makeup from Lancôme (Le Crayon Waterproof Creme EyeColour in Moka, Les Aquatiques Waterproof Creme EyeColour in Tortue on lids, Dore on browbone, Les Aquatiques Waterproof Mascara in Black), Joue à Joue Blushing Creme in Stardust Rose, Hydra-Riche in Fraise de Bois on lips. Mary Jane Marcasiano knit tank. Alexis Kirk earrings. Hair by Didier Malige for Jean-Louis David. Makeup by Rumiko."

Who the model is as a person—her personality, her politics, how she looks when she gets out of the shower—all of this is irrelevant. She is like an artist's canvas; her only role is to take paint well.

Despite appearances to the contrary, a photograph delivers not true knowledge about the object photographed but rather, as Susan Sontag observes, "a semblance of knowledge, a semblance of wisdom; as the act of taking pictures is a semblance of appropriation, a semblance of rape." Models turn out to be only a starting place in any quest to understand the ideal image and its impact on the viewer, about as important as a canvas is in understanding a painting. Indeed, the information they can provide about a photograph in which they appear is of the same order as what a rape victim knows about a stranger who rapes her—informative, yes, but circumscribed.

One must look beyond them, at the process by which the image is made, and the cultural uses to which that image is put. The professionals who have that information work from the *other* side of the camera.

THREE

No Body Is Perfect

Diana is a smart, frisky, pretty thirty-six-year-old photographer whose work appears regularly in national magazines. When I arrived at her huge Greenwich Village studio, she offered me a seat at one of her work tables and handed me a pile of her photos to look at while she finished a phone call. In thumbing through them, I noticed a familiar face. At first I couldn't quite place her, but then I realized it was Mimi—looking quite different.

When I met Mimi at her apartment her hair was flat, her face was free of makeup, and she wore a sweatshirt and jeans. In these prints, Mimi was vibrant and youthful-looking from her hair down to her high heels.

I asked Diana how she made a woman Mimi's age look so smashing. "There's a certain amount of disguise involved," she replied, at the beginning of a long description of what it takes to "package" an older model. "Women over forty start to lose it around the chin, and their necks start to show more age. Discoloration in these areas is a big problem—you notice freckles sometimes, or liver spots. What you do is try to even things out. For instance, if someone has been in the sun, as Mimi had, her body is likely to hold more color than her face. So instead of making the face very light, which is what you do in beauty shots, you match the face color to the skin color. You have to strike a balance and make it all look uniform.

"If someone who's not a professional puts on makeup, you may find that the makeup sits in the wrinkles, and then you get a very clear line when you take a photograph. To avoid that, you have to shoot quickly enough that the makeup doesn't settle into the skin. The makeup artist needs to use just the right amount of makeup and sponge it out and matte it with powder. Then you shoot immediately.

"With an older woman," Diana continued, "you might work with umbrellas on both sides of the face, casting the light very evenly and spreading it over the face. Then you use a reflector, a silver card, underneath, just to throw a little extra light to fill up the eye sockets, the circles under the eyes, and a little bit under the chin. But sometimes you have to make compromises. Mimi has a prominent chin, so we couldn't use the reflectors. We left her with a little line under the eyes instead of emphasizing the chin."

I asked if anything special had to be done below the neck. "The important thing is posing," Diana answered. "An older model may have flabbiness under the arm, but you deemphasize that. If she puts her hands on her hips, and she doesn't have enough musculature through the forearms and upper arms to make it look nice, you have her bring her shoulders back and keep her arms down, or perhaps folded in front. You always want a nice clean line through the waist to the midriff, rather than a bulge, which even the fittest woman at that age may have. You communicate your attention to that spot by having her take a deep breath and stretch up, so you don't have any funny little curves where there shouldn't be any.

"Then there's positioning of the camera. It may be better to shoot somebody straight on if she has a clean line, say the bust through the waist and the hips, than to have her turn around, where she may have a little more bulge over the tummy area, as Mimi has. A lot of women her age also have too much bulge around the hip, so you may have her shift her legs, or shoot from three-quarters height and twist her shoulders back to the camera so that you don't have that wide spread across her hips and thighs."

In other words, even if you're attractive to begin with, as Mimi is, you won't be able to go out to dinner looking like the women in the ads unless you can take along an entourage of makeup artists, specialists in posing, and lighting technicians.

I ended up spending the better part of two days in Diana's tutelage, during which time she dispelled any notion I may have had about the realism of photography. Photographs do not *pre*sent people, they *re*present people. They inevitably exhibit the person from particular points of view.

Even with the most exquisite model, some camouflage is usually called for. Diana showed me an article in *Gentleman's Quarterly* about her favorite, Paulina Porizkova, among the most successful models of the eighties, who made half a million dollars in just one year. *GQ* quoted Paulina as saying: "People just think it's my style to just always be the sadly smiling girl in all the pictures, the dreamy-looking one. I'm dreamy-looking because I can't see the camera without my glasses and closemouthed because my teeth don't come out great in pictures." She's had her teeth fixed since then, but the point Diana wanted to make is that any part of a woman's body that is not softly rounded is felt to inhibit her appeal. So fashion photographers generally direct the eye away from ears, teeth, and elbows, and select models without bony knees and shoulders. And, as Diana explained, you'll almost never see feet in ads because "these beautiful models come in here and they're six feet tall and wear size-ten shoes."

There's no such thing, I came to understand, as the all-purpose body, the one so perfect it needs no improvement and can be used to advantage in any advertising layout. Rather, modeling agencies offer basic categories of perfect bodies to choose from—body types that can be mixed and matched in the service of particular products.

"Two common body types in modeling are the swimmer and the dancer," Diana reported. "The difference is in configuration. A swimmer's body often has very broad shoulders, strong hands, and narrow hips. Dancers are usually longer, more stretched out. Instead of being packed in the shoulders, their musculature travels all the way through their body. They're like cats."

How do you choose a body type when you're casting? Said Diana, "For some products it's specified what kind of girl you need. For lingerie, you have to be sample size—thirty-four B cup, and you have to fill that cup, you can't be less or more, because the bra won't work. And you have to have a thirty-four or thirty-five hip, and long legs and a slim waist. Those girls are more like dancers than swimmers, except they have to have busts.

"There are so many factors to take into account when you are actually casting. The models can be long-waisted or short-waisted and still look good. It depends on the assignment. If you are doing something on waistlines, you are going to look for a dramatic waistline of whatever sort, but if you're doing five pages on exercises, you're going to pick somebody with not just a good waistline but a good hipline too."

The phone rang at that point in the interview. Actually,

it rang frequently during our meeting, with calls from makeup artists, art directors, and the babysitter who was taking care of her two-year-old son. Without fail, Diana would return from intricate negotiations with these people about fees and delivery schedules and the baby's dinner, and immediately pick up right where she left off.

"You really want boobs when you're shooting swimwear," she said. "It sells magazines. The main person you have to please is the manufacturer, and the manufacturer still thinks he wants to see a pinup girl. Which girl you choose at a casting will depend on the particular swimwear. If it has a cutout hipline, you want great thighs and buttocks also."

The phone rang again, and en route to answering it, Diana pointed to a magazine on the floor and suggested I look at it. It was the new issue of *Mademoiselle,* opened to a beach beauty spread.

"There are so many factors," she repeated when she returned, pointing to a bikini ad. "We had a swimsuit shoot last month where they wanted two blondes side by side in every shot. The agency sent a couple of girls who looked almost like twins when they walked through the door. Until we started shooting, that is. They were both very thin, but proportioned differently. One was all legs, and so she looked terribly skinny next to the other girl, who had a mannequin sort of hip thrust and shorter legs. She looked almost hefty next to Miss Legs. Normally she wouldn't, but it was the comparison of the two.

"In the magazine, they came out the same size, as you can see, but that's because we kept shooting the leggy one from behind. All you look at is her great back. We put a lot of makeup on the other one, so your eyes go to her face. We had the girls on the steps going into the pool, and that also helped even out the size."

As Diana commented on the more conventional advertising shots from her portfolio, I was struck with a sense of incongruity. How absurd it is that millions of readers believe (or at least *hope*) that by using a particular skin cream their faces will resemble the professionally lighted, elaborately made-up, and photographically retouched skin of the model in the advertisements. Worse still is the notion that by puffing on the brand of "ultralight" cigarette a model is caressing, it's possible to cull some of her apparent sensuality and vibrancy.

What's Wrong with This Picture?

But ideal images in the media do more than simply encourage us to buy products. The sorts of pictures Diana creates affect how we feel about ourselves and people around us, and may actually prove detrimental to the well-being of some who view them.

According to the results of a recent survey of high school and college students in Indiana, portrayals of the ideal body can influence young women toward anorexia. Compared to their peers, the young women who exhibited symptoms of anorexia were significantly more conscious of and influenced by the media, especially magazine ads.

Beauty ads can also sway a young woman's values. In a Texas study, two groups of high school girls watched fifteen network TV ads. One group was shown beauty ads, while the second watched commercials for other types of products. After the screenings, the girls answered questions about which characteristics they considered most important for a woman to be successful in her career and with men, and what they personally found important in a woman. Those who had watched the beauty ads listed more often characteristics such as "a pretty face" and "a healthy, slim body" as more important than attributes like "intelligence" or "hard-working."

Then there's the secondary damage these images inflict upon women. Men often consider normal women less appealing after exposure to images of beautiful women. In one experiment men were asked to rate, at two separate times, the attractiveness of a group of average-looking women. They appraised them before watching glamorous actresses on television and again afterward. The men rated the real-world women as less attractive after viewing the "perfect women." In another experiment, psychologists showed male college students a series of videotapes and slides of women in erotic poses. Then they asked the men to rate the sexual appeal of their own girlfriends. Some of the men were shown images of beautiful women, while others looked at unattractive women. The results: those shown the beautiful women downrated their girlfriends, while those who had seen unattractive women rated their girlfriends higher.

Such influences even reach young children. Studies in San Francisco and Chicago have discovered that 50 to 80 percent of fourth-grade girls are on diets. When a *Wall Street Journal* reporter interviewed some of them, the girls told him, "Boys expect girls to be perfect and beautiful, and skinny," and "Fat girls aren't like regular girls, they aren't attractive." Meanwhile, a boy in the same class alluded to the pressures on his own gender. Girls, he said, "want to be skinny to attract bigger, stronger men, the handsome, dashing kind."

Among the people I interviewed, those who were unattractive or disfigured were *not* the ones most distressed about their looks; nor were they particularly interested in trying to improve their appearance. Just the opposite. People who already resembled the media idols, people who were fairly attractive or even very attractive worried the most about their looks and worked the hardest to try to look better.

My observation is confirmed by other studies. People with serious bodily damage, such as quadraplegics, fire victims, and the very obese, have been found to be less upset about their appearance than people with relatively insignificant problems like a short leg, minor burns, or a few extra pounds.

Apparently, those who have no chance of resembling the ideal are released from its grip and are free to set other goals for themselves. Their lives, however, are still greatly affected by the body ideal of our culture. They're likely to experience discrimination at school and at work because of their appearance, for instance. Still, the homely are in a good position to avoid the neuroses of the age. Although they could improve their looks somewhat, no amount of jogging or cosmetic surgery would turn them into Fondas or Redfords. Their most viable alternatives are self-acceptance and personal achievement.

"It's Sad When Someone Doesn't Accept Himself"

Take Buddy, for example, who is six-feet-two and 280 flabby pounds. "I know you're not supposed to feel that way, but my weight and my face haven't bothered me," he said. "I'm happy. I like the way I am. As a matter of fact, I get a little angry with people who have a fixation about their looks."

Sitting in the living room of his apartment on the second story of a house in an old section of Greensboro, North Carolina, I asked Buddy what he thought of the models in ads. "When I see Calvin Klein commercials or the like, I say to myself, come on, give me a break, how many people in the world are gonna look good in those things?" He jiggled his belly with his right hand, Santa Claus style. "Am *I* gonna look that good in a pair of designer jeans? I look at the guys in those commercials and I think, 'You dumb chowderhead, you probably have the IQ of an ant.'

"It's easier for me to say that, maybe, than it is for a lot of other people. You see, it's very hard to disguise fat, and so fat people have to be more realistic about themselves. There's all that stuff about wearing vertical stripes to seem thinner, but you can't really hide it. You can pad yourself if you're too thin. A woman can pad her bra. But I can't hide this belly, and I sure as hell can't hide this face," he continued, giving a little tug on one of his oversized ears, in case I hadn't notice how clownlike they seemed.

"Some people have the idea that growing means improving, but it doesn't, it just means changing. I'm never gonna look younger. If I can do something to take care of myself so I don't look like I'm eighty when I'm fifty, I'll do that. But I'm satisfied with myself.

"I think it's sad when someone doesn't accept himself. That's never been a problem for me, maybe because I came from a very poor family and was able to achieve something for myself. Nobody in my family had ever gone to college. Only a couple ever finished high school."

What Buddy has achieved is impressive indeed. After twenty years as a highly regarded elementary-school teacher and part-time director of regional theater productions, he developed a children's television show. The program is now syndicated to fifty markets throughout the country, and he serves as writer, producer, and star.

In all of his work he turns his looks to his advantage. Second-graders complain if they're assigned to another teacher's classroom, having heard from older kids about how Buddy will sometimes come to school in a hobo costume or as an oversized Easter bunny. "Being funny is my way of winning people over," he told me. "When I was a kid, my older brother was the tough one in the family and was always getting into fights. It was a lot easier for me to crack a joke than it was to get in a fight.

"By the same token, I was painfully shy as an adolescent. That's the roughest time. When you're a kid you don't take it all so seriously, and when you're an adult you learn to accept yourself. But when you're sixteen and everybody else is dating and you're left in the corner with every acne-faced or otherwise defective kid in the school, you don't feel like making jokes. I just kind of hibernated my way through high school.

"And then, I remember the first few days at college, paralyzed with terror at all those new people. I started in January because we didn't have enough money for me to go in the fall. My father took his Christmas bonus and sent me. The first few days I was too scared to go into the college commons to eat."

Buddy described several painful incidents from that first semester and then explained how, the next fall, he was saved: "I took a public-speaking class because the teacher had been my English teacher and was somebody I really liked. The first speech we had to do was called 'Introducing Myself to You.' I must have worked fifty hours on it. I wrote down everything I could say about myself.

"The teacher had told us to start with something funny to get the audience's attention. I had the speech all written out, except I couldn't come up with a joke. So when I stood up there"—he rose from his easy chair and stood opposite me—"I still remember the first line that came out of my mouth: 'Let's get one thing out of the way at the beginning. I am very fat.'

"That got a big laugh, and it was like an angel came down and lifted a weight off my shoulders," Buddy said as he sat down again. "Everybody listened to what I had to say. I could communicate with people. I joined the Drama Club, and about the third or fourth play I was in, the director, who was short and fat, taught me that I have a tremendous advantage that other people don't have, which is my size. Leading men are a dime a dozen. Pretty people are everywhere.

"Many years ago, I was in *Man of La Mancha*, where I played the barber. He makes his entrance singing offstage, 'I am a little barber, and I go my merry way.' I got applause every night *before* I came out. Most of the audience knew I was playing that part, and that I wasn't little. Sometimes I play off my size directly, but you have to be careful. It's like spices in cooking; it has to be used sparingly or it takes you over. Pretty soon you're just a fat person, and that's not all I am.

"I have a friend," Buddy added, "who stays fat because he says, 'If I lost weight people wouldn't like me.' That never crosses my mind. I don't really worry about what would happen if I became skinny. It would just be like putting on another set of clothes. There'd be the same person inside them."

Size 14 and Doing Fine

That last observation of Buddy's is something that Melissa, a fourth-year medical student I interviewed in Washington, D.C., has only lately come to see. At twenty-five, she's spent nearly her entire life trying to lose weight in the hope that she might become a different person, one more acceptable to her mother and to men. Only in the past couple of years has she learned to accept her body.

But then, a heavy woman in our society generally does have a harder time than a man who is obese. There's no female counterpart to the "jovial fat man" image that Buddy plays so much to his advantage. An obese woman is typically considered unappealing.

Weight plays a greater role in whether a woman is judged

attractive than it does for a man, according to several studies. As a result, women are more concerned with their weight. In surveys, whether of adolescents or adults, about half of the women say they are unhappy about their weight, compared to anywhere from about one-tenth to one-third of the men.

Women have good reason to fret over their extra pounds, since evidence suggests that discrimination against fat women is more intense than against fat men. For instance, a study of admissions to prestigious colleges discovered that the rejection rate for obese female applicants was three times higher than that for obese male applicants with comparable academic records.

Melissa admitted that she, too, harbors deep-seated prejudices against obese people. "When I see someone who is obviously overweight," she said during one of our meetings at a coffee shop near the George Washington University Hospital, "my immediate thought is, They're lazy, they're slobs, they're depressed, and probably they smoke."

She quickly added that, although that's her first reaction, she knows the stereotype to be inaccurate. "I know *I'm* not like that," explained Melissa. "I have plenty of energy, I'm hardly ever depressed, I work my butt off, and I'm way too paranoid about cancer to smoke. Besides, some of the most accomplished people I've known are fat. My father's a good example. He's a successful economist. And when I did my surgery rotation, one of the best surgeons was also one of the largest women I've ever seen."

At five feet, two inches and 170 pounds, Melissa is far from what the insurance tables say she should be (115 pounds), and also far from her own all-time adult low (135 pounds). In fact she's not all that far from her highest weight ever (185 pounds). Yet she accepts her current weight and has no plans to diet. "When I was seventeen," she explained, "and feeling really gross about being fat, I made a list of

twenty reasons I should lose weight: 'Mom will be happy.
Grandma will be happy. I'll be able to buy clothes I like.
I'll get a boyfriend. I'll be healthier,' and things like that.
I'd lose weight, and none of those things would happen, it
wouldn't solve a thing, and so I'd gain it back.

"I look at myself now . . . well, I've fulfilled all those
wishes. My mother and grandmother are proud of me because
I'm about to be a doctor and I'm engaged to be married.
Richard, my fiancé, is a wonderful person. I'm healthy. My
blood pressure's good, my blood sugar is good, my back's
good. I have nice clothes that I like, and I've taught myself
how to dress in a style that's comfortable and expresses my
personality. The only reason for me to lose weight now is
that society is telling me to, and frankly, that's just not a
good enough reason.

"The big difference in me now versus when I was younger
is not my weight. For a while in high school I was a lot
thinner than I am now, but I didn't feel good about myself,
and I certainly didn't have many dates. Partly, I was too
busy being an honors student, which removed me from a
lot of the other kids in Sacramento. Partly, I think, if you
don't feel attractive you project an image of not being attractive
and not being interested in other people."

What about the other kids in her classes when she was
growing up, I asked. How did she feel about them? "There
was a girl in high school who was just beautiful," she replied
immediately, "with very delicate features and long blond
hair—she was very, very thin. I used to look at her and
think, if there was anyone in the world I could be, she would
be it. It didn't matter to me that she wasn't very smart,
and it turns out she was pregnant before she left high school.
She probably wasn't any happier even then than I was, but
I envied her good figure.

"I also remember," she added, in a discouraged voice,

"feeling relieved when there was someone who was fatter than I was in one of my classes. I'd gaze around the room when I was bored and try to guess how much each person weighed."

After she said that, she stopped talking for a few seconds and twirled a strand of brown hair at the base of her shoulder-length shag. "I wonder if I still do that," she asked herself in a quiet voice.

Noticing my interest in her question, she responded aloud: "I've gotten a lot better about it. I have a true sense of my size now, for instance. I used to picture myself differently depending on my moods. When I was in a bad mood I always pictured myself as a really big person, and when I was feeling good or I'd been exercising I pictured myself as a thin person. There are only thirty-four women in my class, and my first year I used to look around and compare myself to each one of them. One is very, very large, and three or four are about my size, and the rest are within the social range of normal, or thinner than that. But the funny thing was, unless I was in a great mood, I always considered myself to be the heaviest in the class. Most of the time I considered myself to be heavier than even this very heavy woman. I used to wish I could stand next to her and reassure myself that I wasn't the heaviest woman in the class."

That said, Melissa took a long sip of coffee before continuing. "I still compare myself to other women," she allowed. "I think everybody does that to a certain extent. I look at *Vogue* and get jealous. I can admit it. But the way I feel now is, I'm never going to be a size six. It's just not possible for me. It's not who I am, I'd feel silly as a size six. At the same time, I've been up to a size eighteen, and right now I'm a fourteen, and I'd like to be a ten. That's realistic."

She paused again. This time her jaws tightened, and she

said, in a louder voice: "I get angry with the way the media portray women as so thin. It holds up an unrealistic goal. I feel bad for younger people who are contending with that. In my own life, I've reached a point where I'm happy with my weight. Not that I wouldn't be even happier if I lost another ten pounds, but I'm engaged to be married now, I'm finishing med school. I'm not that high school girl who still thinks she needs to look like the girl the boys are ogling."

When did her feelings change?

"It wasn't until my second year in med school that I started projecting an image of 'I'm an interesting person.' And once I did that, I started dating all the time. I was dating four or five men at a time right up until I got serious with Richard. No one I really liked, I was just sort of exploring my abilities to attract men. I'd always thought it would be great to have men interested in me; I'd never had that in high school or college.

"So I made a crusade of picking up attractive men. One time I was at the swimming pool, and there was this very good-looking black man. He started flirting with me, and I gave him my number. And I mean, we went out and had one drink and went straight to bed. I slept with him because he was attractive *and* because he found *me* attractive.

"But I phased out of that pretty quickly, once I proved to myself that I could pick up men. Most of the guys I was sleeping with weren't terribly interesting, and certainly not the type I'd consider settling down with."

Melissa's journey to self-acceptance was lengthy and painful. Along the way, she had to come to terms not only with her own view of herself, and with how she was seen by men, but harder still (and more elusive) with her mother's concerns.

"I remember the day I went on my first diet," she related. "I was twelve, and sitting on the toilet, and I remember looking down and seeing that my thighs looked heavy, and I flipped out. I ran to the basement and said to my mother, 'I'm getting fat.' She should have explained to me that it was just puberty and all the estrogens surging. But my mother, who is a wonderful woman, made a mistake that time and said, 'Well, then, go on a diet.'

"That night I had my first binge. I remember very clearly not being hungry and eating everything in sight because I knew the next day we were going to the pediatrician, and he was going to put me on a diet. From that time until two years ago, I was on a cycle of: diet—*gain back*—diet—*gain back*—diet. From when I was twelve until two years ago, I was always eating either a thousand calories a day or five thousand calories a day. There was hardly ever an in-between period.

"My pattern would be: I'd lose weight rapidly at first, and then it would level off and I'd start feeling deprived. Then when my mother wasn't home I would sneak food."

As she described all this, I found myself getting hungry, even though I'd eaten just before the interview.

Melissa went on to describe her mother's involvement with her dieting. On one level, Melissa enlisted her: "I'd say, 'Oh, Mom, I've got to lose fifteen pounds. Please don't let me eat anything for the next month that's not on my diet.' " At another level, Melissa and her mother were both using Melissa's eating as a means to act out their disputes with one another. "As soon as she would leave the house," Melissa recalled, "I'd started eating things that normally I didn't even like. I don't like chocolate, for instance, but when I was feeling deprived, I'd go straight for it."

In this diet drama, her father's role was that of silent compatriot. He, too, is heavy, unlike Melissa's mother, who is average weight. To this day, Melissa told me, he never has said anything to her about her weight.

Still, Melissa described her father as a distant man, and emphasized several times in our interviews that she always felt close to her mother. "I lived with this extraordinary guilt that I was disappointing my mother," she said, "and I'd cry hysterically in my room at night, worried that my mother would notice which foods were missing. And every week she'd ask me how much weight I lost, and I'd lie. I'd say, 'Two pounds,' when I hadn't lost any, and after a while the two pounds added up, and I'd lose track of how much I was supposed to have lost altogether, and she'd catch me in my lie.

"Most of the time I was dieting for her as much as for me. I never felt she was satisfied with me. She always wanted me to be thin. It was like my special job in the family. When we'd have holiday dinners or a dinner to celebrate somebody's birthday or something, my mother would serve my brothers two servings each and my father three servings, and me half a serving. I was secretly resentful of that, and I would sneak into the kitchen later and pig out when everybody was asleep.

"I remember a few times when I was at college," she went on, a little shame in her voice, "not wanting to go home and not knowing why. It was because I had gone up in weight, and I knew my mother would be disappointed. Whenever I'd lose some weight on a diet, she'd be really excited for me and remark on how terrific I looked. Three months later when I gained it back, she wouldn't say anything, then something would slip out like, 'Maybe this summer you'd like to take up jogging.' "

For all that, Melissa maintained a firm (if tangled) tie to

her mother clear through college. She used to call her three times a week—cross-country from Princeton to Sacramento—and rely upon her advice on almost every important matter. Only in the past year and a half has Melissa broken away emotionally.

She told me of a painful transition that began a few months after she arrived at medical school. She grew intensely anxious over whether she could handle the tremendous work load, and she became concerned that she and her mother might drift apart once she was a doctor. Before long, Melissa was on her most extreme diet yet. For three months she restricted herself to 500 calories a day, and she lost forty pounds. "People complimented me on my weight, but I don't think I looked all that good. I looked tired and cranky and drawn," she reported.

"To be honest," she added sheepishly, "I was bulimic. For a couple of weeks I was even forcing myself to vomit."

In fact, Melissa's experience is not unusual. In a survey of female medical students at Harvard, close to half said they are preoccupied with their weight and engage in binge eating. Fully 15 percent have a history of anorexia or bulimia. According to several other studies of eating disorders in high-achieving young women, an important contributing factor is a fear of surpassing their mothers.

But all Melissa knew was that she was getting too thin for her own good, and that she felt out of control. Her response was to inform her parents, early in her second year, that she was considering dropping out of med school, something her father took upon himself to dissuade her from doing.

Over time, Melissa got herself to a psychiatrist, decided to stay in school, and started changing her relationship to her mother. "Sue, my roommate, convinced me that I had become too dependent on my mom," she said. "I denied it at first, so she said, 'Okay, when was the last time you bought clothes on your own?' And she was right. In high school and college I'd bought almost nothing, and when I did go shopping, it was always with my mom. I enjoyed doing it with her, but I had never bought clothes on my own.

"Sue gave me a copy of *Fat Is a Feminist Issue,* or one of those books, which said, 'Don't wait to get smaller, throw away all the clothes you're hoping someday to fit into.' And it's weird, it really *was* liberating to just go out and buy clothes in my real size, all on my own."

During this same period, Melissa's relationship to her father also changed. "He started asking me questions about his medical problems and trusting my answers," she grinned. "When I was growing up, my father was the one who knew everything, so it's nice when there's something that he doesn't know and he needs my advice."

By the beginning of her third year of med school Melissa was completely off the diet treadmill and, as a result of a "therapy assignment" from her psychiatrist, had begun to look at herself in the mirror while undressed. ("People who aren't fat can't believe that someone could be twenty-three years old before she looked at herself undressed," she noted, "but it's true.") In studying her body from various angles, she concluded, "It wasn't my fat, exactly, that was bothering me. What bothered me—I mean legitimately—were my breasts, which had always been huge. I started developing very early, and I got a lot of teasing from older boys. I remember one day: I was running for freshman-class vice-president and I was hanging up posters after school. Three senior-class boys circled me in the empty hallway and started pretending to grab at my breasts, and they were yelling obscene things."

As she described that event, Melissa at first looked pained, but then a roguish smile came to her lips. "I went grocery shopping with my mother when I was home from college at Thanksgiving, and one of those boys was working there, and he asked me out. I had to laugh. But," she went on, returning to the immediate topic, and wearing a more circumspect expression, "having oversized breasts was a problem when I was an adolescent. When I started dating, boys would go out with me just because they were fascinated with my breasts.

"Besides that, I've never been able to find clothes to fit me right. Swimming, which is the one exercise I truly enjoy, was very difficult. So I talked to a nurse I knew who'd had breast reduction surgery, and she showed me her scars, which weren't so bad, and it just seemed like a perfectly reasonable thing to do."

With the objective detachment of a medical student, she summarized the results: "The surgeon removed about two and a half kilos from each side, total weight ten and a half pounds. I'm pleased I had it done. I can buy normal clothes, I swim a mile every other day now at the university pool. My back no longer hurts."

And there was another positive result, she added. "Richard and I met just before my surgery. We had one date before I went into the hospital. One thing the surgery did was to keep Richard and me from sleeping together right away, because I had all the scars and bandages. We were forced to have six or seven dates to get to know each other. I think that really made our first experience of sex something that was based on a relationship instead of just sex. It made me feel that he cared about me, and I didn't feel embarrassed or sleazy or anything, as I had with the guys I'd been sleeping around with before I met him.

"*Also*," she said, drawing out the word as if she wasn't

sure she wanted to disclose what she was about to say, "also, his reaction to the surgery proved to me that I didn't need to worry about his reactions to my body. He wasn't grossed out by the scars—which I guess he shouldn't be, since he's a medical student—but he didn't even respond to my loss of weight. Everybody else commented that it looked as if I had suddenly lost forty pounds. Richard was much more interested in the medical procedures used, and the complications afterward. I just think he's less appearance-oriented in general."

Which is to say, as much as she's come to terms with her looks, she's nonetheless reassured that her fiancé is not someone she must diet for. "Still," she added, "some days I fantasize that in the six months remaining before my wedding I'll lose thirty pounds and be a beautiful bride."

Does she plan to diet prior to the wedding? "Probably not," said Melissa. "It wouldn't change anything at all if I lost weight. Richard thinks I'm beautiful, and I imagine when I spend some time with my hair and put on a nice dress, I will look beautiful, regardless of whether I'm a six or a sixteen."

Does that mean—I dared to push the point—that she wouldn't mind if she *gained* weight between now and the wedding? "Of course I would," she responded definitively, in a tone that implied she'd already given considerable thought to this possibility. "When I overeat, it's for a reason. Used to be, I would turn to food when I was lonely or bored. Eating would blot out all the feelings. All the negative feelings would be gone, except one, which was feeling bad about eating too much . . . and that became the one feeling I knew how to feel. So that any other negative feeling at all I transformed into feeling bad about my weight. It used to be that if I did badly on a test, instead of saying, 'I'm not smart' or 'I didn't work hard enough,' I would say, 'Oh my God, I'm so heavy, I've gained so much weight.'

"I don't do that anymore. The other day, I suddenly found myself staring in the mirror and trying on all my clothes and saying, 'Oh God, this is too tight,' so I sat myself down and thought out what was *really* bothering me. It was that my applications for residency weren't complete."

Melissa is one of the fortunate ones. She has mostly released herself from the ideal images and come to accept her body. Thanks to her strong professional identity and impending marriage, she's been able to build a body image for herself that is based less on society's needs and expectations, and those of her mother, and more on her own.

How each of us looks and how we feel about our looks derive from our relationships with key people in our lives—mothers, fathers, and intimate partners, especially. It's from within our personal dramas that we look out at the ideal images of the media and decide what to make of them. Whether we find those images laughable or take them so seriously that we starve ourselves into anorexia or exercise so excessively that we end up in the hospital depends largely on the messages we're hearing from the important people in our lives.

Sometimes the communication is direct, as when a mother tells her daughter she should lose some weight. But often, the messages are muffled. Many times we're responding to hopes or obligations placed upon us as children, or to the unfulfilled aspirations our parents (and even grandparents) held for *themselves* when they were younger.

PART II

MAKING THE BODY

FOUR

Mothers and Daughters

Ideal bodies are manufactured in studios by teams of technicians. Regular bodies are made elsewhere—in families, for the most part. Statistically, the wealthier our family, the taller, thinner, stronger, healthier, and more attractive we are likely to be. Our parents' background also influences how we look. In the United States, Jewish, Italian, and black mothers on the average raise children who will become fatter adults than children of WASP or Japanese mothers, for instance.

Genetics surely plays a role in this process. About 40 percent of children with one obese parent become obese adults. A whopping 70 percent of children both of whose parents are heavy turn out the same way. It's not all in the genes, though: adopted children of the obese are also very likely to put on pounds.

Whether a *particular* child ends up fat or thin (or beautiful or plain, or delicate or athletic, or lithe or awkward) will depend primarily on what goes on inside the family—how the parents feel about themselves, what expectations they hold for their children, and how everyone gets along. Just as social and historical forces shape the larger world of which the family is a part, within the privacy of the home, family dynamics chisel away at our bodies.

Anorexia nervosa is a dramatic case in point of how social trends and parenting styles work together to fashion people's bodies. Anorexia was quite rare until the past decade, but according to current estimates, perhaps one percent of young women now in their teens and twenties have been afflicted by the disease. As many as one-third of women college students exhibit anorexic-like behaviors such as the use of laxatives, vomiting, and diuretics to lose weight. The disorder is a by-product of a particular family pattern that developed in upper-middle-class homes during the 1950s and 1960s when today's anorexics were children.

Anorexics frequently come from families in which the father was off building a career while the mother stayed at home for the sake of marriage and motherhood. Well-educated themselves, the mothers were frustrated, dissatisfied, and confused about what they were doing with their lives. Toward their daughters they tended to be overinvolved or controlling. Along with their ambitious husbands, they put great pressure on their daughters to achieve, which meant to be pretty and to do well in school.

In such families, a daughter's anorexia is at once an unconscious acceptance of her parents' wishes (she takes the ideal of thinness and self-control to its limit), and a rejection of them (she makes her body boyish rather than womanly).

Ninety percent of anorexics are women. It's difficult to think of disorders among men that parallel eating disorders in the degree to which they distort the body and place the person at risk of serious injury or death. One candidate would be what some psychiatrists have lately termed "macho personality disorders"—for example, the so-called Man's Man, who is ultramasculine and enthralled by rough sports, big machines, and his own muscles. Psychiatrist Leonard Glass has found that the Man's Man is typically the son of a mother who exerted great power over him, so that as a boy he feared he could not break away from her. His father was no help, since he was distant, ineffective, or unpleasant. As a result, the Man's Man repudiates anything in himself that might be considered feminine and tries to keep women, whom he finds perplexing and dangerous, at a distance. In so doing, he implicitly rejects both his mother and his father.

Generally, though, women's bodies are more influenced by their relationships to their parents than are men's. This is because most parenting is done by mothers, who, as sociologist Nancy Chodorow has noted, "tend to experience their daughters as more like, and continuous with, themselves" than they do their sons.

How a woman looks and how she feels about how she looks depend to a great degree on how she has responded to her mother and, in particular, to her mother's body and what that body symbolized. If a woman's mother was self-assured and attractive, the woman may have felt more confident in her own body while she was growing up. Even if one year she was roly-poly with baby fat, only to turn thin and gawky the next, still she could have faith that she was on the road to something better. If, on the other hand, her mother was unappealing—or good-looking but insecure and competitive—the daughter's own liabilities probably seemed more important and permanent. She may well have built her sense of who she is around her extra pounds or spotty face.

Women don't always respond to their mothers by losing or gaining inordinate amounts of weight. Nonetheless, most daughters have no choice but to devise some sort of emotional alliance with their mothers, an alliance that plays itself out throughout their lives and is often reflected in their bodies. The terrain of possible alliances is vast—from those in which

mother and daughter are nearly merged, to mutual antago-
nism—and each has its own characteristic features.

Daughters Who *Mustn't* Look Like Their Mothers

Some girls spurn their mothers, who seem pitiable, and identify with their fathers, obviously the more powerful of the
parents. These women's adult lives are mirrors of their mothers' concerns, but mirrors of the sort you find in amusement
parks. Where their mothers were weak and dependent, these
women try to be strong and independent. Like their mothers,
though, they may be depressed. In their middle years, they
sometimes have neither bodies nor love relationships they
feel happy with or in charge of.

What they may have in their favor, I've frequently found,
is great energy, creativity, and vision. I'm thinking, for instance, of Sharon, a forty-one-year-old professor of English
literature whom I interviewed in a crowded student restaurant
near the University of Texas campus in Austin. Insulated
by loud rock music and ignored by the scampering nineteen-
year-olds around us, we sat at a little table in a back corner,
where we were both so engrossed in Sharon's narrative that
we went through several pitchers of beer and plates of nachos
in three hours.

Sharon told her mother's story this way: "She died about
ten years ago, and she died, as I see it, from a history of
neglect that began with my father and then extended to male
physicians. She had migraines and various kinds of hysterical
symptoms ever since she got married. She went to a psychiatrist, and my father was very scornful of that. He'd say she
was a hypochondriac or going through menopause. The shrink
just gave her tranquilizers. She was addicted to Librium.

"As I and my sisters got older and left the home, her life

contracted dramatically. He moved her away from Bay Shore, where we grew up, to a house in a new development close to Kennedy Airport. It was much more convenient for him, since he flew all the time on business.

"So my mother had very few friends, no community, no children. Her world became a television set. She gained weight, she began not to do much of anything, and she deteriorated. Nobody ever took her seriously. She was bleeding anally for six months. They kept telling her, 'It's just hemorrhoids.' She had to have a colostomy, and she was so weakened that she had a stroke."

As her mother's death became imminent, Sharon's own life began to change. She was thirty, in a bad marriage, and for a dozen years had been dropping in and out of colleges. She left her husband, quickly finished her undergraduate degree, and enrolled in a Ph.D. program in English. "I took to literature like a fish to water," said Sharon, a short, oval-faced, animated woman with bushy red hair. "When my mother died, things just reconfigured for me, and suddenly I felt the necessity to get my Ph.D., to get a full-time job, to move out of the range where what happened to her could happen to me.

"At that same point I started to reassess my father's role in my life. Until then I'd identified with his point of view and thought it was just silly the way my mother acted. I'd seen him as the intelligent and active one, and I'd been scornful of my mother's passivity. He used to travel around the world and bring back gifts for everybody in the neighborhood. My mother lay around the house doing nothing."

The overriding fact about Sharon as a child was that she had been overweight—a fact that played into the family dynamics.

"I always had the message from my father that my weight

was a real problem and that it was important for me to get it under control. I remember my mother and father quarreling about it. My mother basically adopted the policy, 'Let her do what she wants, it's baby fat, she'll grow out of it,' but my father said, 'No, it has to be gotten in control.' He was terribly manipulative, but at the time it must have seemed to me that he was the one who cared about me and my health.

"Yet there was no pleasing him," she went on. "He expected me and my sisters to excel as women *and* as men. He wanted us to be superfeminine and superattractive, but he also expected us to be unafraid and independent, and he was unforgiving of any feminine weaknesses. If we'd show any emotion he'd become angry."

In response to these conflicting demands from her father, Sharon copied the only woman she'd seen him interact with. Like her mother, she was moody and self-pitying, and attracted Dad's attention by being sick. She also tried to keep her weight down and her femininity up, but her unconscious objection to her father's tyranny and her mother's wasting away made that effort impossible.

Sociologist Lucy Rose Fischer, who studied a sample of mothers and their grown daughters for ten years, suggests that all daughters, including those who as children were overtly distant from or rejecting of their mothers, continue in adulthood to reproduce some of the key features of their mothers' lives.

Early in the interview Sharon described her mother as "a classic nineteenth-century hysteric living in the twentieth century." Late in our meeting, the nachos all gone, Sharon revealed how she herself is split between the centuries: by day she lives in the thoroughly contemporary world of the radical feminist college professor, sporting big abstract-design earrings and teaching at a campus built on oil money; by night she lives in a hundred-year-old house with a man named William, the spindly fifty-year-old owner of an antiquarian bookshop.

Sharon and William have, she said, "a lot of intellectual communication." But she turns elsewhere for sex. "It's like Virginia Woolf and Leonard Woolf," explained Sharon. "Not that I'm comparing myself to Virginia Woolf, but as long as I keep my other relationships in their place and I come home, that's all that matters."

However literary she tries to be about the arrangement, though, it's not an easy one to maintain, and there was a recent episode in which she'd neglected to keep her sex life sufficiently "in place." Half a year before our interview, Sharon broke up with a man ten years her junior when he insisted she leave William. The results of that breakup, she said, were a weight gain of twenty pounds and a greater commitment to William.

As she talked about her affairs, it became evident that her figure, too, is caught between the nineteenth and twentieth centuries. When she's not in a sexual relationship but solely with William, she's the epitome of the full-bodied Victorian woman, but when she's pursuing someone else— usually a man younger than she, and very good-looking— she's constantly exercising and on a diet, and considerably thinner.

Sharon and William live together in the closest thing to marriage either feels capable of. It's a safe relationship because it's free of all the dirty stuff that goes on between husbands and wives—like sex and babies and jealousies—and because it's ennobled by a set of abstract beliefs. And Sharon's lovers fill in for what she doesn't have with William. Still, the package doesn't work to calm the unease she feels, which fuels her brilliant and highly acclaimed scholarly writing but also causes periodic rages directed at William.

Sharon described William as "the father I never had," but actually her relationship with him is more a continuation of her confounded feelings about her *mother*. When a woman remains, in Nancy Chodorow's words, "ambivalently dependent on her mother, or preoccupied internally with the question of whether she is separate or not," she's likely to develop a relationship with her husband (or husband substitute) in which she's simultaneously dependent and angry.

The only place Sharon does feel reliably powerful is in the classroom—where she's flying high and talking business, like Dad when he was away from Mom. "There's an incredible transformation when I'm lecturing," she said. "I feel very rooted and in charge, and very attractive, and I *never ever* feel fat, even when I'm way above what I should be. Teaching a class, I never feel vulnerable or exposed or as if I should be cowering. My body stance is very different from what it is when I'm on the street or at home. I'll let myself do very masculine things sometimes in class. I'll stand with my leg up, I'll talk in a tough manner. I'm in charge there, I'm setting the standards for what is attractive. I'm the creator of a universe."

Daughters Who Embody Their Mothers' Dreams

That image of Sharon in class popped into my mind some months later, when I was interviewing Juliet, a famous theater and film actress. She, too, has been driven since early childhood by the desire to make her life different from her mother's. For her, too, performing is a type of liberation.

Unlike Sharon, however, Juliet is happy with herself and

her marriage. Her story differs from Sharon's in several ways. Rather than repudiating her mother's fragility and negativism, Juliet has always been her mother's secret ally, and she has found a way to bring her conflicts into her work in a redemptive manner.

Juliet decided at age five that she would go into the theater. In seventh grade she came upon an article in *Seventeen* about the American Academy of Dramatic Arts and vowed to attend. She talked about the fantasy constantly with her mother, who pooh-poohed it as unrealistic. Nevertheless, Juliet saved enough money from after-school jobs to take the bus to New York for the audition, where she was accepted.

"Once I got to New York," she recalled, "I became a different human being. I'd felt trapped in Iowa, and I'd gone into my own little shell. In New York I was more outgoing and articulate, and a whole lot prettier.

"Everyone I knew in Iowa felt trapped. Both of my parents had sacrificed their dreams. My mother was a nightclub singer when my father met her, and my father wanted to be a football coach. Both of them came from poor families, and there wasn't money to go to college or start any kind of career. They felt constantly isolated and burdened. My father held three jobs most of the time I was growing up, so my brother and sister and I almost never saw him."

Juliet's mother was not much more accessible. Except to cook dinner or slave over the washboard, she hid in her bedroom, felled by her nerves or a virus. She was also prone to unpredictable fits of anger, in which she'd throw things at the wall, terrifying Juliet, whose role in the family was to reassure her siblings that Mom would be okay again, and to tell Dad not to worry, everything would work out.

Periodically her mother left the house for several days. One time she was gone for three weeks, and the whole family assumed she'd never come back. She's since told Juliet that she had simply camped in the woods and cried or sang to herself until she felt she could face her life again.

Her mother's troubles always registered physically as well as emotionally. When Juliet was five, her mother had what doctors at the time referred to as a "nervous breakdown." "That was the worst episode," said Juliet in a doleful voice, "but I can remember many times seeing her body go into that sort of shocked state. Every few months she'd sink into a feeling of 'I don't know what to do,' and then she'd run away or get sick. It continues to this day."

It was one day when her mother was in the hospital that five-year-old Juliet was home from school with the flu. Alone in the house, she ventured onto her mother's bed, where she found a movie magazine. At that moment Juliet fixed on the idea of becoming a grand and beautiful actress far away from Iowa City. Yet it's fair to say (and Juliet says as much herself), that in virtually every role she plays, she's exploring the anguish her mother experienced but couldn't share, and portraying the women her mother might have been had some of her dreams come true.

"I felt that I had a vision, that I didn't know exactly what that vision was, but that somehow I would tell the story of these people and try to understand them," said Juliet. "A few years ago, when I was rehearsing a character on Broadway, something happened. I had learned the lines and was working on the movement, and a voice came out that wasn't mine. I remember feeling a heavy weight in my body. Wardrobe was going to give me extra padding so I'd look heavy, but they didn't have to, my body just dropped. I became square-looking, squatter. My shoulders fell right toward the middle.

"It was frightening. This play was about a woman who felt trapped and isolated. Every night before the performance, I would just collapse into a state of pain and sadness. After the performance I'd gradually come out of it, and the next morning I was myself again, but every night I would live her life from beginning to end for two and half hours on that stage."

Juliet described this to me at nine in the morning one cold day in March. We sat at a small table beside a red brick fireplace in the living room of the Upper East Side Manhattan brownstone she shares with her husband. The last bit of a log from an early-morning fire was expiring.

Before I met Juliet, I'd expected a large, tough, menacing woman. In the roles that have made her famous—a sadistic nurse, a murderous wife—that's the way she looked. Instead, I found her very attractive, svelte, and soft-spoken.

At my request, she described what it's like to inhabit the body and mind of two different types of women—dejected women like her mother and the openly angry and strong-willed women she so often plays. "That second type of woman gets into a passion more," Juliet explained. "Her passion keeps her alive, whereas women like those I grew up with lost their passion or had their passion denied them. You feel exhilarated after a performance when you play one of the angry ones."

The interplay of dreams and passions is at the heart of Juliet's acting technique. Many actors develop a character first through the body, because bodily movements, unlike thoughts, are visible and reproducible. Juliet is just as physical in her approach but uses the mind to learn about the body, rather than the other way around.

"I find my way into a character through her hopes and dreams. What did she imagine for herself, and where did she get shot down? When did reality come in and say, 'No, you're not going to get your dream, it's going to remain just a dream the rest of your life'? How did she accept that, and is she still dreaming? What happiness does she have in her life, and where are the sorrows? Those are the things that determine how she looks and carries herself."

In her own life, Juliet recently fulfilled a dream that she had long ago abandoned: at the age of forty-six, she got married.

"Now," she said, "I am aware of aspects of womanhood I never knew before. At this age in my life, a lot of the parts I get are mothers and wives. Now I'm able to show the tension they feel, because I feel it at a very deep level. There's an anger that's always there, and the question is whether you're going to turn it around or let it take you over and weigh you down. When I play that anger on the stage, it's a mixture of gentleness and harshness, of private fears that something could happen to this cherished relationship, and at the same time the need to hang on to my own independence."

Daughters Who Do Their Mothers Proud

Daughters of mothers who have come to terms with, or even relished, their roles as wife and mother tend to inhabit their bodies quite differently. Instead of fighting their mother's agony, they carry forward her positive self-image. These daughters build their own identities on the strong foundations their mothers pass along to them.

Often it's in minority families that one finds the sort of mother I have in mind. Sociologist Lena Wright Myers reported that two-thirds of the black women she studied had wanted, as teenagers, to be like their mothers. Now that they were adults and mothers themselves, they told Myers, they still admire their mothers more than any other women they've known.

Many black women, like many immigrant women, devote their adult lives to building a better future for their children. Take, for instance, Katherine's mother, who grew up in rural Louisiana, on the banks of the Mississippi. She married when she was eighteen, and she and her husband relocated north to inner-city Baltimore in hopes of a better life for their children. Katherine is eternally grateful to her mother, whom she credits with transforming her from a poor, plain little girl into a prosperous and beautiful woman.

She remembers her mother as a taskmaster. "When we were growing up my mother constantly urged us to 'e-nun-ci-ate' our words. Any grammatical error we made we'd hear about. She wanted us to speak *English*. Her sister was really funny about that: usually she'd say 'axe' instead of 'ask,' but as soon as company came over she'd start sounding like an English aristocrat. I had a hard time when I was first in school because some black people equate our family's style of talking with being stuck-up."

It was also important to her mother, a tidy, heavyset woman, that Katherine look right and behave properly. "When we would be driving to visit friends of the family, Mama would say, 'I don't care what they offer you, don't you dare take anything.' If we were impolite my father would snap his fingers and my mother would give us a look. We'd straighten up real fast."

The tightrope between guiding a child and dominating her is a difficult one to walk, expecially for a woman who holds a definite vision of how she'd like her daughter to turn out. The consequences of falling very far to one side or the other are great. Studies show that adult children of mothers who were either overcontrolling or underinvolved often hold negative images of their own bodies as adults. Katherine's mother was neither too much nor too little involved. In fact, she verged on being overbearing only in the matter of boys. "It used to bother me because I always did really well in school, good grades and cheerleader, honor society. But when it came to dating, she was a prude. I couldn't go to a boy's house, stuff like that. But there was a lot of love that went along with it. My mother would never just tell me to do something. She would sit down and explain it to me."

While her mother was director and drama coach to the five children, Katherine's father played producer. He was a traditionalist who allowed only his sons to have bicycles or play sports, but he saw education as inherently good and pushed all his kids to do well in school. "He can't wait for me to get my law degree," Katherine announced proudly. "He's already picked out his suit for graduation."

She was not considered very attractive in high school. "I thought I was fat, because everybody else thought I was fat. Now that I look back on it, I think a lot of it had to do with going to an all-girls school and being shapely at thirteen. A lot of my friends were still really skinny and didn't have any chest or hips. But I had the kind of body that forty-year-old men would stare at.

"I thought I was bigger than I really was. When I was fourteen and we got our cheerleader uniforms, people got upset when they found out that mine was the same size as theirs. They thought of me as fat. My sisters used to kid me that I was the fat one."

The mixed message worked to Katherine's advantage. She was officially fat and brainy and so no sexual threat to friends, older sisters, or mother and father, and she was sanctioned to shine in school. At the same time, her peers had to recognize that she was actually as thin as they were, except that she was developed.

The result of the collaborative efforts of Katherine and her mother is the Katherine I interviewed in the lobby of the Boston University Law School. She fit our talks into her packed schedule—which included classes, *Law Review*, study groups, a part-time job in the legal department of a Fortune 500 corporation, and an active dating calendar.

To look at her you'd think she must have been raised by well-placed WASPs. At our meeting she was in a yellow sweater, square yellow earrings, stylish black cotton slacks, and yellow socks—strictly preppy pre-professional. She wore her hair permed, which she said gives her flexibility. She can keep it compressed and simple at work, add body and movement for school, and even spike it a bit if she's going to the clubs to dance.

Katherine knows how to look good to both whites and blacks. When in the corporate world, she dresses primly, in simple stylish clothes that only hint at her figure. When she wants to look sexy on a date, she pulls herself into a tight skirt or jeans that emphasize her bottom.

"My white friends," she said, "have boyfriends who like them to be very skinny. But black guys can't stand slim women. I think it's what you grew up around. If your mamma is skinny you like skinny women. But it is acceptable for black women to get fat as they get older. All my aunts and my mother are really shapely. Big hips, big thighs, big butts.

"When I go home to Baltimore, some guys on the street might see me from the front and not say anything. They have to see me from behind. Once I get past them they start screaming at me, 'Nice ass!' Black guys think a big butt is really attractive.

"But I have a hard time fitting into business suits. They're made for a white woman's figure. In the business world people are sensitive to appearance: you can't be overweight—black women are just larger than white women, but if you're going to be a lawyer, you have to watch your weight—and you have to look nice, but if you look *too* nice, too put-together, people think there's nothing between the ears. I'm very conscious of that."

Katherine's life is a struggle for balance, whether she's talking about her relationship with her mother, or how she comports herself, or which male admirer to go out with on Friday night. "I grew up around all black people, and it's always been important to me to know where I came from," she said. "My family talked about roots before there was *Roots*. I have a lot of white friends now, but I never want to be one of those black people who become successful and never go home. My best friend is the girl who lived next door when I was growing up. We always will be best friends, even though we're different in a lot of ways. She got married, and she and her husband live at her folks' house. We talk twice a week on the phone, and I see her whenever I go home.

"I know people who have themselves dropped off a block away from where their parents live because they don't want their white friends to see they live there. I never want to get that way. I know how to go out with white people and drink with them and feel comfortable with them. But I think you want to keep in mind who you are and where you come from."

Few of us are as competent and comfortable at bridging diverse worlds as Katherine is. Most choose one to the exclusion of others. People who have made it out of the ghetto and into the suburbs typically want to have little to do with members of their own group who are still poor. To maintain dual affiliations requires great social skills. Permission to use those skills usually comes initially from parents. If a mother gives off signals that by succeeding in the larger world her daughter is deserting her, her daughter will likely hold back. If a mother is ashamed of her own background, her daughter might try to leave it behind, only to suffer confusion over her identity. In either of these scenarios, a daughter has cause to grow plump and matronly in symbolic affiliation with her mother and her other female kin.

Katherine and her mother have the rare sort of bond between two family members—one sees it also in good marriages—in which both parties are committed to and rely upon one another emotionally but lead independent lives. A great danger for a daughter of a strong mother is that the daughter will lose her own individuality. Katherine escaped this pitfall because her mother encouraged her to be her own person and to choose her own career.

On the other hand, if a mother is *merely* a role model, someone to copy and become, her daughter may be in for trouble. If the daughter doesn't have the basic looks or intellectual ability to follow in her mother's footsteps, she'll likely spend her life fending off feelings of inadequacy. On the other hand, if she *does* she faces another risk. She may become a living replica of her mother and never forge a separate identity.

Daughters Who Look Like Their Mothers

More than a small number of daughters resemble their mothers—and vice versa. In our youth-adoring society, where the ideal image of womanhood calls for perpetual late adolescence, this is hardly surprising.

There's even a nationally televised tribute to bigenerational twinning—The Mother-Daughter Pageant. During the 1987 pageant, one of the finalist moms, in her question-answer session, said of her daughter: "The first thing that comes to mind when I think of Tammy is how similar we are in our values and our thoughts and just the way we do things. So often I look at Tammy and I feel it's like looking in a mirror." Indeed it was, and Tammy had her own little story. "On my birthday," she told the host, "I bought this outfit especially to impress my parents when they came down to take me out to dinner from college, and Mom *had the same outfit on!*"

Out in the real world, it's usually difficult to be your mother's twin. *As* difficult, in fact, as being her opposite, like Sharon, and more dangerous, too. What becomes of you when your mother shows signs of aging, or becomes seriously ill? Or worse yet, what happens when she dies? How can you follow in her footsteps then?

Those are exactly the dilemmas that confronted Erica, thirty-nine, who had been designated at birth to be heir to the family femininity, mother's girl. In important ways, she's benefited from that inheritance. She's very attractive and entirely comfortable with her body—a statement I would make about very few people I've met.

In studies of attractiveness, social psychologists have found that beautiful little girls receive more attention from parents and teachers, who consider them more appealing, smarter, and better behaved than their siblings or peers. Later on, beautiful women typically end up with wealthier husbands. And all of this has come true for Erica. The beauty and grace her mother passed on to her have served her well at several crossroads in her life. But at the same time, she suffers great confusion and disappointment about who she is and where she's going, especially when she tries to move away from the role she learned from her mother. She's fine when her principal activities are attracting men and raising her beautiful daughter, but when she tries to fashion a career, things don't seem to work out.

Our interview took place at her and her husband's grand, newly renovated Victorian house on the outskirts of Hartford. We sat in the living room, where I began, as usual, with questions about her childhood. "I was the pretty, quiet one, and I loved to play with dolls and pretend I was the mother," she reported. "My younger sister was supposed to be a boy but turned out to be a girl. My mother used to tell people that all the time—they got their girl with me, but their second child was supposed to be a boy. She became a baseball-playing tomboy who always beat me up."

Erica walked across the room to pick up a photograph of her mother from the mantel. In the picture, her mother looked exactly like the Erica sitting before me. Erica had done everything with her mother, she told me. They would bake and sew, and comb each other's hair. Her mother even led Erica's Girl Scout troop. The long-range plan was that Erica would train at a local college to become a nurse, as her mother had originally intended to do. But when the nursing program turned her down, she and her mother had to find an alternative. "The obvious one was teaching, but I wasn't a great student, so what would I teach? I thought, 'Why not phys ed?' Well . . . everybody thought it was a riot that I should decide to do that when my sister had been the jock."

So Erica stole a little of her sister's thunder. Not so much, mind you, that she was ever in danger of forsaking her feminine birthright. While her sister captained the women's basketball team at the University of North Carolina to first place in the region, Erica swam and played tennis at Virginia Tech, and not well enough to make a varsity team. She did manage to win the romantic attention of the most sought-after fraternity men at the university, as well as a place on the synchronized swimming team. From synchronized swimming it was a short glide to modern dance, which became her passion.

Late in Erica's sophomore year, just when everything had fallen into place, her mother, who was only forty-six years old and seemingly in perfect health, died of a heart attack. Erica went through a massive identity crisis. "From the very start I'd been raised to *be* her. Now she was gone. I felt as if I had to step into her shoes immediately. I went home and took care of my father for a couple of weeks while my sister, who was openly lesbian by then and excommunicated from the family, stayed in Chapel Hill. The whole time I was there I was extremely restless."

Upon her return to college, she promptly broke up with the varsity linebacker she was pinned to and started dating a guy named David, "an intellectual loner, different from anybody I had ever met. He listened to classical music and read poems by Ezra Pound."

A moving force behind her interest in this scrawny Jewish philosophy major, who was as close to a hippie as they came at Virginia Tech in 1967, was a need to remake her life, to break out of her mother's biography, to become someone else entirely. Erica started wearing jeans and no makeup, and lashing out at the other young replicas of mothers she found all around her. "I lived in a sorority house at the time. I remember having fits that the girls were so silly and immature," she remembered. "David's grandfather was a Communist, and he'd send pamphlets to David every week. I'd devour them and go back to the sorority and lecture the girls about the plight of blacks and anything else I could remember."

Erica and David married during spring break their junior year—and, soon after, Erica's life began to replicate her mother's. She and her husband found jobs together in a progressive high school in Vermont, where Erica taught dance and David taught social studies. She became pregnant two years later and quit work. Among the *Mother Earth News* crowd they'd befriended, mothering was all the rage. They ate natural foods and breast-fed and vowed to raise children who would wage international peace.

Nonetheless, full-time baby care is full-time baby care no matter which ideology it's dressed up in, and without her job Erica had few opportunities to dance. Once the thrill of pregnancy and early motherhood passed, and winter set in, she became depressed, argued with David a lot, and gained five pounds. This was the first of two times in her life she'd gained weight, and she said it terrified her, because her mother had always been slender. Even so, when she'd look in the mirror, Erica told me, she'd see her mom and start to weep. She missed her horribly.

It was Freud who observed rather dryly that, "under the influence of a woman's becoming a mother herself, an identification with her own mother may be revived." Such identification brings closer together many mothers and daughters who have drifted apart over the years. But in the case of a daughter who had always been closely identified with her mother, and who lost her without so much as a good-bye, the feelings that new motherhood brought were difficult to cope with.

Erica grew increasingly dejected and edgy. When summer came, she left her daughter with David and took the bus to New York City for a week's vacation with a woman friend from college.

Her friend introduced her to someone who ran an experimental dance group on the Upper West Side. "I just danced all over New York, day and night," she recalled, her big blue eyes open wide. The trip got extended to two weeks, then three. When she finally returned home, "there was perpetual conflict. The people I'd met in New York were not in the same situation I was. I had a husband and a child and lived in the country. But in my heart I was a dancer.

"The conflict was never quite resolved," she said broken-heartedly. "In a certain way it was resolved in the shattering of that marriage over the course of the next couple of years."

So that Erica could work, the couple moved to Hartford, where Erica began a string of affairs, one with an eighteen-year-old, another with a visiting French teacher, and several others that also couldn't last. "It was a time similar to when I went away to college," she recalled. "I felt validated by being attractive and feeling that I was hot stuff. Yet I felt awful about myself because I didn't want to be playing around."

She was eventually yanked out of her marriage by a self-confident man her own age, who wasn't going away, and who wanted more than her body. She described this new man in words precisely opposite to those she used to describe David. Where David was "weak, gentle, soft," Hank, her second husband, is "forceful, strong, and sometimes nasty." Neither extreme is fulfilling. "Sometimes," disclosed Erica, who has been married to Hank for five years, "I wish I could crawl back into the coziness of my first marriage."

But one of Hank's endearing traits, she said, is his generosity. Before they were married, he offered to pay her way to a Ph.D. in education at the University of Connecticut. Burned out on teaching sweaty sixteen-year-old girls, she took him up on the offer. But when she received her degree, she found that jobs at her level were scarce in the Hartford area, where Hank owns a thriving restaurant supply business. She's bored, and dismayed at the thought that she might have to return to teaching after all the work she put into getting her degree.

What's more, for the second time in her life, her psychological state is adversely affecting her appearance. She's gained five pounds, which she said she takes to be a disturbing sign that her days as a dancer are numbered.

As happened with her mother at the same age, the brightest spot in Erica's life at the time we met was her daughter, Chloe. When I commented that Chloe, whose picture was also propped on the mantel, looked a lot like Erica, she agreed. "We're very close. We have a dance class together when she gets home from school today.

"I'll tell you something that may sound strange. I knew she was a girl before she was born. I had chosen the name Chloe before I got pregnant, so I must have had some premonition. I feel I really knew her even before she was born."

Much as a woman can pass along to her daughter the physical attributes she inherited from her mother, so too can she pass on her overidentification with her mother.

Given the enormous influence of mothers on their daughters, it would seem that a girl has no choice but to grow up in her mother's shoes. The only question appears to be in which way she does. She can adopt her mother's form of femininity, as Erica did (and pass it on to her own daughter in turn); or adapt it, as Juliet and Katherine did; or repudiate it at her own peril, as Sharon has.

But the story doesn't end there.

FIVE

Father, Food, Fitness

Early in the nineteenth century, Alexis de Tocqueville observed in *Democracy in America* that "American women, who are often manly in their intelligence and in their energy, usually preserve great delicacy of personal appearance and always have the manners of women, though they sometimes show the minds and hearts of men." Tocqueville was especially taken by frontierswomen, who were expected to be tough survivors out in the wilds, and nurturant and frail at home.

From our earliest years, these two ideals of American womanhood have been at odds with one another; and yet if a woman wishes to be viewed favorably, she must embody *both*. On the one hand, she's required to be the good wife—gentle, caring, dependent; on the other, she's expected to be a good worker (whether inside or outside the home)—rational, strong, and capable of managing complex logistics and relationships.

A woman who will not or cannot adjust to these incompatible demands runs the risk of becoming emotionally disturbed. In the nineteenth century, one of the most common female disorders was hysteria, whose key symptoms were violent outbursts and a childlike need to be taken care of—behaviors at once a caricature of and a protest against the two poles of "proper" womanhood.

You don't hear much about hysteria these days. Times change and so do the symptoms people use to express their conflicts. Now we have eating disorders and exercise addiction, two conditions which at first glance may seem very different. Anorexics and bulimics scale down their bodies, after all, while exercise obsessives build theirs up; the first course of action destroys health, while the second, at least on the face of it, improves it.

Yet eating pathologies and exercise pathologies are closely related. Whether a woman is strung out over calories or over repetitions on a Nautilus machine, she's engaged in a furious battle to purify and perfect herself. In either instance she's striking out against the limitations of occupying a traditionally female body. Susan Bordo, a philosopher, has pointed out several similarities among anorexics, bulimics, and exercise junkies: all fear "the takeover of the body by disgusting, womanish fat," stop themselves from menstruating, and yearn for absolute control over their bodies. A study published in *The New England Journal of Medicine* found additional parallels: both anorexics and exercise fanatics tend toward depression, hold high expectations for themselves, don't openly express feelings of anger, and are preoccupied with food and with staying thin.

In part, women who suffer from eating disorders and exercise addiction are responding to their mothers. In the traditional view, the woman who starves herself or vomits after she eats is reacting to her mother's arbitrariness and inability to relate to her. Another interpretation, argued brilliantly by Kim Chernin, holds that a young woman's eating disorder is an expression of her troubled but profound *bond* with her mother. "Is it any wonder," asks Chernin, "that so many women are tempted to take on lean, male bodies in the hope that they might escape from the mother's destiny without enduring all that remorse" of rejecting her overtly? And in the same vein, one might ask whether women who build up their bodies through exercise are motivated in part by a wish to avoid their mother's vulnerabilities or subservient role.

Yet the woman with an exercise addiction or eating disorder is also reacting, frantically and unconsciously, to her *father*. "We think bulimia is surfacing now," say psychiatry professors Susan and Orland Wooley, "not only because young women are under great pressure to be thin, but because they're pressured to be 'strong' in other ways. For the first time in memory, young women are expected to grow up to be more like their fathers than like their mothers. Those new expectations can make the transition from adolescence to adulthood excruciating, and bulimia can become a way of dealing with the pain." Fathers of bulimics are usually traditional, independent, successful, and emotionally distant men, the Wooleys report. "To become like their fathers, our patients feel compelled to be thin—not just to minimize their womanliness, but also because thinness, in this culture, is a sign of achievement and mastery. The bulimic woman's thin body proclaims that she is as strong and lean as a man."

We've seen all the crucial meanings that a mother's body holds for her daughter. But the father's body symbolizes a great deal as well. Isn't the image of father the one Susie Orbach describes when she writes of the anorexic: "Although she looks extremely frail, she feels herself to be strong, to have defeated the exigencies of the body, to have overcome its human limits. Her 70-pound body can run eight miles a day and work out for hours on the exercise machines. She doesn't need food, she doesn't need to respond to the unseemly appetites of the ordinary female body. While others consider her pathetic and in need of help, her self-image is one of which she is proud. She feels strong and impenetrable."

In important regards, the anorexic is the counterpart of the macho man; both put on a tough, slick exterior to cover up secret fears about their own competence. The anorexic "submits her body to rigorous discipline to divert her from her feelings," Orbach reports. "She experiences her emotional life, particularly her needs and conflicts, as an attack on herself and thus she attempts to control these feelings so that she will not be devoured by them." The tragedy is that unlike her male counterpart, whose pathology may actually help him make it in the world, the anorexic literally withers away.

Many anorexics (and bulimics) have gone on to become exercise fanatics. Therapists sometimes recommend exercise as a treatment for eating disorders, and not entirely without justification. The fate of the exercise-obsessive is less bleak than that of the anorexic. She may be able to continue her obsession for years and feel energetic and confident in her work and relationships. But she exposes herself to physical danger as surely as the woman who refuses to eat. Athletic injuries resulting from overuse and misuse can result in long-lasting, or even permanent damage to bones and nerves.

As healthy as they look, exercise addicts are suffering—physically and psychologically. A study of obsessive joggers found that they use running as their major coping mechanism for countering depression, stress, and unresolved psychological conflicts, and that they feel devastated if forced to quit running for just two weeks.

Several exercise addicts have gone public in the past few years. Judith Elman, a San Diego psychotherapist, used to run every day, sixty miles a week, and would even jog with her patients ("to gain freer access to the unconscious, thereby facilitating more rapid progress in therapy"). After severe damage to her knees, which she made worse by ignoring the problem, she had to give up running. "The first months without my daily fix were characterized by almost unbearable tension, the feeling that some invisible force was ripping at my gut, a volcano about to erupt any minute," Elman wrote in an article about her experiences. She became depressed and lost interest in her work. Soon she was catching colds, strep throat, bladder infections, and finally, pneumonia.

It's not easy for an exercise addict to cut back. "There are no programs or rehab centers to detox us from our fixes of sweat," notes Blair Sabol, a writer and self-confessed "exerholic." "We're the only addicts for whom exercise isn't a therapeutic way back."

In Elman's case, many months passed before she recovered from her misery through counseling and the use of visualization, a meditation technique in which one imagines the painful parts of the body repairing themselves. "Perhaps the greatest benefit I gained from visualization was a sense of control," Elman reports. "Control over one's life—the ability to make choices—is vital for a positive self-image and a feeling of personal power. And control, I realized, had been a crucial ingredient in my addiction to running. Through running, I could control everything from my weight to feelings of guilt, anger, sorrow, and fear. But now I had come to see that running was only a prop; the true sources of control were my own heart and mind."

Control. If I had a dollar for every time I've heard an exercise or thinness buff use the word, I could buy a health club.

Control is the postwar obsession. A study of articles in women's magazines from the 1940s through the 1970s found that self-control—not love, friendship, beauty, or wealth—was the most frequently mentioned topic.

Probably the basic reason so many of us are fixated on self-control is that we're constantly reminded of how little control we have. Every night on the news we hear about terrorists and rapists, accidents and AIDS. Cancer, the great killer of our age in whose shadow nearly everyone stands terrified, may still strike even if we give up smoking and take in more fiber.

In an immediate sense, though, our cravings for control are not so much the result of a fear of disease or even death, but rather a fear of *vulnerability*. Just look at the popular media in which we express our collective paranoias. Horror films of the past dozen or so years have come to rely for their terror less upon the possibility that the victims will die than upon the certainty that awful things will happen to their bodies. Take the moment in *Alien* when the creature springs forth from a crew member's stomach. Or that unforgettable warehouse of bodies suspended in midair in *Coma*, slated to be used for spare parts.

Women's bodies have traditionally been viewed as being in special danger of going out of control. Because women give birth, they have been equated with nature and its unpredictability; and women have long been imagined to be ruled by their feelings.

Historically, the job of teaching women to keep their bodies in check fell to men, ranging from religious leaders to corset makers. But during the nineteenth century, mothers were the agents charged by preachers, doctors, and political leaders with instilling discipline and a fear of excessive emotionality in their daughters. Without such restraints, it was feared, a young woman might fall prey to hysteria, promiscuity, or worse.

Well into the twentieth century, mothers took charge of inculcating self-control in their daughters. All this changed after World War II, when a new philosophy of child-rearing caught on. Its manifesto was Dr. Benjamin Spock's *Common Sense Book of Baby and Child Care*, which has sold more than 30 million copies. Parents abandoned traditional notions, which emphasized firm rules, firm punishments, and little overt affection, and treated their children as individuals to be cuddled and enjoyed.

Millions of Americans from white-collar families—now in their twenties, thirties, and forties—were raised by mothers who believed that their major job was not to contain their children but to promote their free expression. Having grown up without imposed control, but in a world where control is highly valued, many desperately seek it as adults. And more to the point of our current discussion, they seek control in masculinized ways. The model of self-control in the postwar family was Dad. Dad enforced whatever rules *did* exist; and Dad was typically the more rational and unemotional parent. From a child's perspective, he symbolized, as Signe Hammer has suggested, "Don't Worry, Everything's Under Control."

Papa's Big Girl

Many modern daughters allied themselves with their fathers when they were growing up, sometimes with unfortunate results later in life. Carrie, a thirty-two-year-old Seattle high school teacher who used to be anorexic and is currently an exercise addict and health-food buff, recalled: "If I would offer an opinion at the dinner table, it would be accepted by my father. If my younger sister would say something he would say it was stupid, so that eventually she just shut up."

Being heard at the dinner table was one of the benefits of being Papa's girl. Prettier than her sister and a better student, Carrie was neither so pretty nor so clever that she was a threat to her father, a hard-driving Jewish wholesaler of men's apparel. Carrie was "Papa's little mensch," as he still refers to her.

Research has shown that fathers without sons often treat their oldest daughters as "substitute sons." Carrie's father played ball with her and put her to work in his warehouse from the age of eight, leaving Carrie's sister, Debbie, at home with their mother, who was usually depressed. Debbie, in turn, developed stomach problems. As an adult she has been plagued by viruses and respiratory problems.

"I'm just the opposite. I'm never sick, never take medicine," said Carrie. "People who meet us can't believe we're from the same family. She's short and heavy and unattractive. I'm tall and thin and"—she laughed—"well, I'd like to be a lot better-looking, but certainly I'm not *un*attractive."

Carrie would have you believe that her sister is the big loser, but there's no hiding the fact that Carrie is just as uncomfortable with her physical self. The two of them simply express their unease in opposite ways. Debbie is overwhelmed with how her body has let her down, while Carrie works hard to prevent her body from getting the best of her.

Carrie keeps as far away as she possibly can from Debbie, who lives in Florida; in fact, Carrie's body is an emblem of how far removed they are from one another. "I'm real concerned," she said firmly, "that I maintain a healthy body. I never eat meat, and I keep fit. I don't *like* it when I'm sick or I feel fat." She hasn't missed work for an illness in over two years, and at five feet, seven inches and 113 pounds, she's thinner than the current ideal. That's because her rigorous exercise program keeps her middle tucked in, she's flat-chested, and in keeping with the old bromide, she *thinks* thin. "I've always wanted to be skinny, and I still do," Carrie told me during our conversation in her immaculate kitchen. "Not skinnier than I am now, that's unhealthy, but I've learned to accept the fact that I have a basically anorexic mentality."

She frets over every speck of fat or sugar that might have slipped into her diet, and her exercise regimen at the health club after work each day involves a three-mile run, a game of racquetball (she recently won third place in a citywide tournament), and for dessert, an advanced aerobics dance class rigorous enough to exhaust a dockworker. In summer, when she's not teaching and can devote herself full-time to exercise, she increases the running to four or five miles outdoors, and she adds competitive swimming and long-distance biking.

When I asked what she gets out of all that activity, Carrie replied with a purposeful look on her face. "I like the way I look and the way I feel. I've never been particularly happy that I don't have breasts, but ninety-nine percent of the time now I love being flat, because as far as exercise goes, it's an asset. Sometimes in clothes I wish I had more shape, but basically I look pretty good compared to most Americans, who are not as fit as they should be.

"I feel that I can control my life and how I feel, which is a nice feeling," she added.

And a rare feeling in Carrie's biography. As a child, her father controlled virtually every aspect of her life, from what she did after school to when she started dating. When she went off to college she immediately sprang loose—drinking liquor, smoking dope, and having sex while under the influence of both.

"I was like a bird flying for the first time," Carrie said,

her cheeks reddening a little. "I felt independent but scared. I had never been on my own or been allowed to make decisions. I never even liked alcohol, but here I was drinking and getting stoned three times a day."

She went cold turkey off drugs and alcohol when she returned home for the summer and promptly sank into a depression that didn't lift until she went on a diet, losing twenty pounds before the start of school even though she hadn't been overweight.

By pulling herself out of her depression in this way, Carrie escaped being "nothing but a bundle of nervous symptoms," like her mother. By dieting she refused her mother's food. She also cemented her bond with her father by deciding to major in math. Her diet turned into full-fledged anorexia that fall semester back at school. When her weight dipped below ninety pounds and she still complained about looking fat, her roommate took her to the student health center.

"I honestly didn't feel there was anything wrong with me," she remembered. "I was just doing what other girls were doing, only better. They all wanted to be thin, I was superthin. Everytime they'd lose a few pounds, they'd gain it right back, but I just kept losing more and more. The doctor told me I was emaciated, but that's not the way I felt. To tell you the truth, I've never felt better in my life. I was running five miles a day and hardly eating anything. I felt totally in control of my own body."

She said this in a puffed-up voice, as if she were still not sure she wanted to abandon that glorified version of her disorder.

"The psychiatrist," she went on, "told me I was doing this because I didn't want to have sex with my boyfriend. He may have been right, in that I certainly *wasn't* having sex. But the odd thing was I've never in my life felt more passionate. Not toward men, but within myself. I felt that I was really alive and beautiful, that I was becoming a whole new person, someone I could admire."

She was also becoming a person at risk of serious hormonal, cardiac, and gastrointestinal problems, a fact her psychiatrist finally convinced her of, which served to shift the focus of her obsession from food onto athletics. By junior year she was devoting a good percentage of her time to the women's track team, and in her senior year when she met Bob, who is now her husband, her weight had actually risen to a record 115. "I was overeating to make up for all the undereating I'd done, but I rationalized it by saying that the only way to stop being an anorexic is to eat, eat, eat," she said, acknowledging the yo-yo effect that's so common in the lives of the overcontrolled.

It's hard to avoid the conclusion that Carrie found herself another father figure in Bob, a man fourteen years older than she who had supervised her during her student teaching. "I really don't remember feeling attracted to him," she confided. "I was just in awe that he was such a good teacher. The attraction came more from personality than from anything physical. He didn't fit my image of the all-American boy I was going to marry."

She did marry him, four years before our interview, and both say they're happy together. Bob, a bearded fellow who wears flannel shirts and down vests, has been a jock since elementary school. He is in great physical shape and is not put off by Carrie's fitness zealotry. In fact, he told me, this is his third marriage, and he doesn't mind that Carrie is so busy with her work and exercise that they get together only to eat, sleep, or make love. After his first wife, who demanded total attention, and his second, who was habitually unfaithful, he appreciates the freedom to go off with his buddies after work and know what Carrie's doing and that she won't be upset if he returns late.

Besides which, Bob admits to a weakness for young women, and between Carrie's devotion to exercise and the clothes she wears (corduroy jeans and turtleneck sweaters from the teens section of J. C. Penney), she looks nineteen years old.

From Carrie's vantage point, these preferences of Bob's mean she's appreciated. The four years of their marriage have been the happiest, and certainly the most settled, of her life. And yet, on some level, her fears about herself have not been calmed. "I fantasize having affairs with someone closer to my own age," she revealed at one point in the interview, in the context of describing a daydream she had the previous week while at home nursing a racquetball injury. "I'm very attracted to Bob, but in my head I hear myself saying, 'Well, he's going to die before me, and then things will be out of my control. I'll be old and not attractive to other guys.' I go through this whole scenario in my mind when I'm feeling insecure. It's silly, because Bob is probably healthier than ninety percent of the men my age. He eats right and keeps in shape."

What's more, there's a practical matter on her mind. Carrie has toyed with the idea of having children, preferably before she turns thirty-five. Yet neither she nor Bob has been able to make the commitment. Bob hesitates because of the two kids from his previous marriage: one refuses to speak to him and the other lives a couple of thousand miles to the east with her mother. As for Carrie, she imagines pregnancy and motherhood not as a gain but as a loss.

"Having a child should be this wonderful maternal thing, and yet here I am, very selfish," Carrie disclosed. "I just think about my body the whole time, because I know when

I don't feel fit physically it really affects me emotionally. And I don't want to resent this kid if I can't get back into the kind of athletic shape I'm in now, or if something should go wrong and I'm unable to do things in the future."

To give up athletics, even temporarily, is a scary proposition for Carrie, especially since it would place her squarely in her mother and sister's camp. To some extent, probably every woman experiences motherhood as giving away her body and turning herself into food for another, thereby diminishing herself. For a woman with a history of eating disorders, this can be a frightening prospect indeed.

When I asked whether cutting back on exercise might not be such a bad thing, given her own description of herself as compulsive, Carrie snapped: "Ten years ago I was anorexic. Five years ago I was bingeing and gaining weight and feeling fat. I used to talk all the time about how I had to get in control of my body. 'I look fat,' 'I shouldn't be eating all this junk.' In the past several years, instead of just talking about it, I'm taking positive action."

Carrie's response pinpoints what it is about exercise that makes it so appealing that thousands of people have become addicted to it. The more they perspire, the cleaner they feel.

Said Carrie: "I like to sweat. When I finish playing a game of raquetball it looks like I came out of the shower. Women are not supposed to sweat like that, but I do. I just think it's a very healthy cleansing system to be able to sweat and then replenish your fluids with a long drink of water."

The Polluted Papa

But why would women feel in need of purification in the first place? Partly because they got the message that they were dirty from their mothers when they began to masturbate, menstruate, or date. Partly because the culture at large still defines women in impossibly dichotomous terms—pretty or ugly, pure or sullied, fat or thin.

But fathers play an enormous role as well. Susan and Orland Wooley, in their work with bulimics, discovered that fully half of their patients had been sexually abused, commonly by their fathers or father substitutes. A sense of shame contributes to their hatred of their own bodies, the Wooleys report: "For relief, they turn to self-purification: they avoid food and they exercise obsessively, trying almost to demolish their bodies." In other cases, it's not a direct assault by her father, but the negative connotations of what his body *meant*, that stand between a grown daughter and her contentment with herself, her sexuality, and her relationships with men. As much as a father's strength or worldly accomplishment may inspire a modern daughter to emulate him, so can his *weakness* become a daughter's albatross.

When I interviewed Lynn, an executive secretary in Indianapolis, and one of an estimated 11 million daughters of alcoholics, I could feel her father's presence in many of her answers. His alcoholism played on Lynn's psyche, not because he was abusive (in fact, she remembers him as kind and warm) but simply because he was an impossible burden on her and her mother when Lynn was a child.

Early in the interview, I asked Lynn about the types of men she finds attractive. "Men who are physically fit," she responded, "not necessarily *built*, but physically fit, who obviously take care of their bodies. Healthy. Overweight men turn me off, guys who smoke turn me off." The man she's engaged to fits her requirements, she hastened to add, and he's tall, well-groomed, and on the fast track in the accounting firm where he works.

She hasn't always gone for that type of guy. After her father killed himself when she was twelve, she found herself attracted to dirty delinquent types whom she tried to rehabilitate. In high school she went steady with a boy well on the road to alcoholism himself. After he totaled his father's car, she presented him with a melodramatic bluff—"It's me or the alcohol"—and he promptly dumped her.

Studies show that daughters of alcoholics are more likely than other women to marry alcoholics and to become depressed after doing so, and Lynn's biography is true to form. Her first husband, whom she married just after graduation, was in some ways a considerable improvement over the desperados she'd been dating. He was ambitious, and his problems with booze didn't emerge until after she'd left him. During their marriage, he rented a small storefront, where he built and sold kitchen cabinets at night and on weekends after his regular job packing boxes at a warehouse.

Lynn encouraged him in his carpentry and shared his dream that someday it would turn into a full-time business that could raise their standard of living. She grew dismayed, though, that he spent all his time there; after the first few months of marriage, they almost never had sex. "At first I blamed myself. I felt fat and ugly. I kept buying new clothes, because none of mine seemed to fit right. Which is ridiculous, because I look exactly the same now as I did then, and I feel prettier now than I have in a long time."

Over time, Lynn sank into a depression. She watched TV, slept a lot, gained weight, and seldom left the house except to go to work or shop for groceries. "I got into the kind of state my mother used to," she said, referring to the gloomy years before her father's death.

But, like her mother, Lynn is no quitter. She kept on fighting with her husband and trying to interest him in her. One Friday night, after a scratching, yelling, plate-breaking argument over why they never went out, Lynn gave up. She phoned a single woman friend to take her out.

She made herself up, and she and her friend toured the singles bars until two in the morning. In Indianapolis, a young woman like Lynn, with a pleasing face and busty figure, can reconfirm her sex appeal rather quickly. A half-dozen men tried to buy her drinks and ask her out.

Lynn's husband, who'd always seemed harmless if nothing else, threw a fit when she returned home. Unbeknownst to her, he had been taking amphetamines at work and had popped a couple extra when he came home from the carpentry shop and discovered his wife missing. He hit her twice, hard, on the jaw. She ran out of the house and drove to her mother's apartment across town, where she spent a sleepless night.

The following morning, while her husband was at work, Lynn called in sick to her boss, packed, and moved her belongings to the home of the friend she'd gone bar-hopping with.

That afternoon, instead of collapsing or seeing a lawyer or counselor, she drove to her old high school and, with a gym teacher's permission, practiced gymnastics for a few hours. The better part of the following two weeks, which she took off from work, were spent on trampolines, balance beams, and rings. When the equipment was occupied by the students, she ran at the school track in a bright red jogging suit and New Balance running shoes she bought herself. "It made me feel that I had some control over my life," she remembered. "It made me feel that I was holding things together. I had no idea what I was going to do with my life. Or actually, I *knew*, but I wasn't facing up to the fact that I had to leave my husband for good and start over, and I didn't feel very attractive. I don't mean attractive *outside*, this had to do with something inside."

Adult offspring of alcoholics often respond to troubles in

their lives by engaging in addictive behaviors, psychologists have found. Common alternatives are eating disorders and drug abuse, but Lynn chose exercise. She had excelled at gymnastics in junior high school and returned at this point to a familiar source of comfort and control—the one interest that had kept her going after her father died.

What a guilty liberation his death had been. She was confused at first—at age twelve it's not easy to comprehend how suicide could be the logical next step for the son of a Dutch immigrant laborer who knows he's worth more to his family dead than alive. But now that she didn't have to go directly home after school to take care of her father, she was able to test out an interest she'd been developing in gymnastics.

"It was something that not a lot of kids were doing," she said, "and I got very good at it very quickly. I won prizes while I was still in junior high. It made the other kids think more of me as a person."

So thrilled was she at feeling good about herself that before long she was tackling the beams before and after school, on Saturdays and Sundays, and performing calisthenics secretly in her room at night when she was supposed to be studying. By age eighteen she'd ripped certain ligaments so many times that an orthopedic surgeon recommended installation of an artificial hip. Lynn and her mother refused, correctly it would seem, since her problems are confined to some bursitis. ("It bothers me mostly when my legs are up during sex," she said, laughing.)

Despite the pain, Lynn never gave up gymnastics, except for that year of her marriage. In the seven years since she left her first husband she's kept herself in shape with aerobic dance, gymnastics, and jogging. She confesses to a great deal of pain at times and says she's accepted that at age twenty-eight she can no longer perform most of her old gymnastics routines. She's even gotten to the point that she can miss a day or two of exercise without feeling awful about herself. But ask her if all the pain is really worth it, and you see how committed she is to the fitness ideology: "A lot of people don't take the time to stay physically fit," she said with conviction, "and I consider myself smarter because I'm taking care of myself. You get one shot at this protoplasm you're stuck in, so you might as well keep it in the best shape possible. I don't have a lot of respect for people who don't take care of themselves."

Her main conflict, she told me, is not over the pain but over the time involved in her athletic pursuits. Her job is leading nowhere, and she'll always be typing letters and pacifying clients unless she gets a college degree. So she'd like to take courses after work. "This year's out. I've got a wedding to plan," she said. "As it is, I hardly see my fiancé because I have to stay late at work so often. Luckily, he works long hours himself and doesn't mind. But I'm always trying to cram twenty pounds of potatoes into a five-pound bag."

Why doesn't she substitute a math or English course for one of her gymnastics or aerobics classes? I asked.

"When I'm not exercising," she replied, in a tone that implied she's thought about this question, "my mood is not elevated. I think there's a definite correlation between the times I'm not exercising and when I'm depressed. I know that if I want to be happy or continue being happy I will continue to exercise."

According to therapists, children of alcoholics urgently seek to feel positive about themselves and in control, having blamed themselves as children for their parents' drinking and having lived with so much uncertainty around the house. Through great discipline, Lynn has achieved both self-esteem and a sense of control in the last couple of years. One sees it in the prudent way she dresses, in the fact that her weight hasn't varied by more than a pound and a half on her electronic bathroom scale, and in the sober, fastidious man she's marrying.

Who's Got Control?

The passion for control, felt by so many Americans in recent years, has its roots not only in individual family pathologies but in the conditions of the larger society. The types of control people exert over their own bodies correspond to the controls that their culture imposes on them in other spheres of their lives. "Bodily control is an expression of social control," is the way anthropologist Mary Douglas expressed it.

Douglas observed that hair, clothing, diet, and sexuality tend to be highly constrained in such historical periods as recessions and wars and in places like blue-collar neighborhoods and military barracks, in which external forces exert great power over the population. On the other hand, in times and locations of relative prosperity and reduced control by the state, people experiment with their bodies and relax the curbs they place upon themselves.

Examples come easily to mind. The long hair, bra-shedding, and sexual freedom of the 1960s took place at the same time that major institutions of American society, particularly the family, began to exert less direct control over young middle-class Americans. On the other hand, the exercise fads and conservative styles of dress during the last dozen or so years were accompanied by increasingly restrictive social controls in the larger society. During Reagan's reign, conventional religion and marriage made comebacks; and mergers, job shortages, and high housing costs convinced many young Americans to defer to those in power.

That women are subject to greater external control than men is undeniable. Notwithstanding innumerable articles in the media suggesting that political and economic equality for women is close at hand, American women are in fact still ruled by men. In 1988, women make up only 2 percent of the Senate, 5 percent of the House, and two out of fifty governors. Recent political history includes the defeat of the Equal Rights Amendment and the first woman vice-presidential candidate. Conditions may appear considerably better for women on the economic front, but only until one looks closely. In the mid-1960s, women constituted only about one-third of the labor force, compared to almost half in the mid-1980s; and the percentage of women managers and professionals has nearly doubled. But despite this influx, women haven't been accepted on equal terms with men. From the mid-sixties to the mid-eighties, the share of clerical jobs held by women has increased from 70 to 80 percent. And working women generally make about two-thirds what their male counterparts do, little better than they did two decades ago. What's more, women are still responsible for about 80 percent of the housework and child care, surveys show, even when they have jobs.

Today, women's obsession with self-control, and in particular with exercise, may have everything to do with false expectations—with being brought up to take their place in society alongside men, and then being denied the power and privilege that men enjoy. If, as physical beings, we are attracted to exercise because it makes us feel clean and healthy, as *social* creatures, we're enticed by the sensation of power. But the "fit" female body of the 1980s is, as Wendy Chapkis has noted, a symbol of power without much real authority.

Lynn and Carrie have something in common besides their devotion to exercise. Each has had a great deal of responsibility in her family, and in some cases at work, but because of the idiosyncrasies of their parents and partners, these women have never been truly *in charge*. So they redirect their need for control inward.

This impression was bolstered when I met Christine, a woman who *is* in charge and always has been. Her aversion to exercise is as strong as Carrie's and Lynn's attraction to it.

At forty, Christine serves as senior vice-president for a Chicago-based Fortune 500 company. Her massive executive office is exquisitely decorated with an oversized desk and sleek Scandinavian sofas, one of which she directed me to as she sank into an easy chair across the way. There was an odd feel to the office: it was at once sophisticated and sentimental. For decoration, in addition to original paintings by prominent contemporary artists, around the room were several girlhood mementos: a doll, a teddy bear, and a grade-school picture of herself.

Christine's own appearance also seemed something of a paradox at first. Although she boasted that she never does sit-ups and she eats whatever she wants, she had the slender figure and beautiful, youthful face of an actress.

The incongruity was quickly resolved, as she told me of her love affair with cosmetic surgery. When her thighs started to spread a few years ago she had a suctional lipidectomy performed. When she noticed bags under her eyes last year, she had them tucked. Her executive health insurance package covered both operations.

Christine is the very embodiment of the ideal image that advertisers and health clubs purvey, but minus the "no pain, no gain" ideology commonly associated with the image. Christine wants everything that goes with the good life, including the beautiful body *and* the ceaseless comfort. Of her suctional lipidectomy she said: "I just found that every time I looked at myself in the mirror, I would be cupping my thighs with my hands to see what I would look like without saddlebags. So I made an appointment with a plastic surgeon and found out it was completely safe and simple.

"I would have had to do ten thousand leg lifts to get the effect of that painless little operation, and still I would have had some flab, just because I'm getting older. I had it done just before my birthday, so my friends came to visit in the hospital and brought me presents. It was fun. When I got home, the man I was seeing at the time was very solicitous and wouldn't let me pick up anything or really do anything for myself for the whole recovery week. I got to stay in bed and watch old TV shows."

Christine is boss to thirty-five managers, who in turn manage 300 employees. The only kind of control she wants after she leaves the office is the remote kind that flips the TV channels. "I don't cook, I stop off at a restaurant on my walk home," she said. "I have three regular places, they all know which table I like and how I like my favorite dish prepared. When I get home I take off my clothes and get into bed under my nice soft sheets with my teddy bears. If there's nothing good on TV, I watch something I taped the night before. That satisfies me on every level.

"You see, I don't do things that don't give me pleasure," she added in a slightly lofty tone of voice. "That's been true my whole life. I have a wonderful life, and I'm sure from some people's point of view it's lazy and stultifying and minimizing, but I like it. I'd hate to ever have to give it up."

When she describes her childhood, one can appreciate why she'd surround her adult office with reminders of that period. Her parents, who were fairly well off, made sure she was comfortable and happy. Both were busy, her father at work and her mother in community organizations, but they frequently included Christine in their activities and consulted with her on such matters as where to go for vacation or which new car to buy.

She didn't do especially well in high school or college, as she was busy with dates and cross-country skiing and European trips with her folks. Yet she has succeeded in the business world, thanks to ambition and a knack for being in the right place at the right time. She started out as a secretary in the corporation where she now works. When her boss left the company, his successor recognized that Christine, more than anyone, possessed the knowledge he needed to do his job well, so he offered to make her his executive assistant if she'd stay. In that post she developed the skills and contacts throughout the building that allowed her to move into the marketing division, and from there, a couple of years later, to her current title.

Christine's three mottos in life, she made sure to tell me, are "Discretion is the better part of valor" (a paraphrase of Shakespeare), "Abstinence is easier than moderation" (a paraphrase of St. Augustine), and "Whenever possible, take the easy way out" (source unknown). She applies these to her work by taking on only those assignments she knows she'll do well, and never revealing her professional or personal weaknesses at the office, however tempting that may be.

She also applies these precepts in her approach to her body. She never eats desserts, because once she starts she has to have one with every meal, and she doesn't exercise, because cosmetic surgery is an easier way for her to maintain her good looks. "There's some discomfort each time you have surgery," she admitted, "but frankly, it's a lot of fun compared to going to the gym and working out for two hours a day, four days a week for as long as I live."

Looking to the future, Christine said, "I've been talking with my surgeon about my derriere, but I think I'll wait a few years. It could be a nice forty-fifth birthday present to myself."

Why *would* Christine exercise? The three major inducements to women to exercise—health, beauty, and power—don't apply to her. She's healthy, and she figures that the walking she does to and from the office exceeds the minimum weekly allowance for aerobic activity. She has her good looks, both inherited and purchased. And as for power, she's got plenty.

Were Christine employed in a corporation where it is the norm to be seen at a health club, or where in order to conduct business one must play golf or tennis or run laps with clients or partners, matters might be different.

But male executives are more commonly under pressure to prove their physical fitness. Usually, as long as a woman keeps herself trim and energetic, she has met her obligations. Indeed, if she's too athletic, her male colleagues may find her intimidating.

Exercise and control continue to carry very different meanings for women from those for men. If for some women the health club is a way station on the road to equal citizenship and social power, for their male counterparts it's an old and familiar workhouse. Athletic ability and muscular development have long been routine obligations imposed on every American male.

SIX

Men and Muscles

America was built of male muscle, at least according to our popular lore. The standard version of our early years speaks of rugged pioneers fighting the forces of nature and mastering savages with their bare hands. American industry likewise is understood to have been the product of male brawn. The captains of industry in the late nineteenth and early twentieth centuries were portrayed as almost animalistic in their physical power and drive. Aspiring young men were urged to display their own commitment to the same values. A 1920s manual for salesmen, like some of its counterparts in the 1980s, recommended exercises each morning, because muscular strength "imparts a feeling of enthusiasm, physical vigor and power of decision that no other faculty can give."

Bernarr Macfadden, creator of the physical culture movement early in this century, exhorted men to realize that "it lies with you, whether you shall be a strong virile animal . . . or a miserable little crawling worm."

During the world wars, male strength was equated—in political speeches and posters—with patriotism. And men who grew up just after World War II remember vividly the Charles Atlas ads in comic books of the period. "I manufacture weaklings into MEN," read the headline on the back page of a 1952 issue of *The Fighting Leathernecks* (ten cents a copy). Beside a huge picture of Atlas, "the world's most perfectly developed man," appeared the famous story of how he used to be a ninety-seven-pound weakling. The choice every man had to face is made explicit in these ads: he could either keep his "skinny, pepless, second-rate body" or turn it over to Atlas (or the high school coach or the trainer at the local gym), who would "cram it so full of handsome, healthy, bulging new muscle that your friends will grow bug-eyed."

Generations of boys have received the message loud and clear. Sociologist James Coleman asked high school boys in the early sixties how they would like to be remembered. Nearly half chose "athletic star," far more than opted for "brilliant student" or even "most popular." Neither hippies nor drugs nor the women's movement has changed things very much since. When the same question was asked of high schoolers in the seventies and again in the eighties, the same results were obtained: close to half answered "athletic star." What's more, in contemporary studies of college students, muscular men have been shown to be better liked by others and happier with themselves than their less well-developed classmates.

Boys suffer if they can't or won't accept the obligation to develop manly physiques. Every one of 256 nonmuscular adolescent boys examined in one study suffered mood or behavior problems connected to feelings of physical inadequacy. *Sissy* is, after all, a much more negative term than *tomboy*. While a girl is expected to outgrow her tomboyism, a boy who doesn't act boyish may well be sent to a psychiatrist for help. So a boy must prove decisively his commitment to masculinity, and the primary way to do it is through athletics and muscularity.

Muscles are *the* sign of masculinity. Author Nancy Huston has pointed out that women are distinguished from men by their ability to give birth, but men have no parallel "mark" of their gender. To fill in for this lack of a distinctive male trait, Huston says, many cultures have granted physical strength to boys and men as a characteristic uniquely their own. Over the years, innumerable scientific and superstitious explanations have been advanced purporting to prove it was God or Nature that made males stronger than females.

Because of the great meaning attached to muscles, nonathletic boys often grow into insecure men. In an "About Men" column in *The New York Times Magazine*, Mark Goodson, the television producer, wrote humorously about the drawbacks of disliking sports. Soon after arriving in New York in the 1940s, "hungry, anxious, in need of work," he was offered a job hosting a sports quiz. "I felt the blood leave my face," he recalls, but he accepted the assignment. Every Monday night for twenty-six weeks he feigned an interest in the subject, well enough that the radio station offered him a job announcing a baseball game. Never having been to a baseball game, he rushed out to buy a book on the rules of the game. "As I got to the tenth page, I collapsed," he reports. "Much as I needed the money, I knew there was no way that I could manage this bluff."

Goodson built a TV production empire despite such setbacks, and he jokes about them now. But he also recognizes that to be male and nonathletic is serious business. "I approach this subject with a light touch, but in truth," he writes, "it has been a problem that has plagued me for most of my life." From early childhood until late adulthood, he hid his disinterest and inability for fear of seeming homosexual. Yet "even after three marriages, three children, and some in-between love affairs, plus the sure knowledge that I adore women, I still feel, from time to time, that, somehow, I must be lacking in the right male genes."

One irony, of course, is that for many years now it has not been much easier for a man to be nonathletic if he's gay than if he's straight. The ideal man within the gay world, as in the heterosexual, is powerfully built. "What a shock I had when I came out," said Jim, a twenty-nine-year-old real estate agent I interviewed at the San Francisco apartment he shares with his lover.

Jim had waited until his junior year in college to become involved in the gay community. One aspect of coming out that he'd eagerly anticipated was the opportunity to dress the way he wanted. As far back as he could remember he'd been careful not to wear flamboyant clothes and to camouflage his thin arms and concave chest with a sports coat or sweater.

"I was basically a sissy as a kid," he told me, "and I had a lot of defenses about it. I would get stomachaches from having to play baseball. The whole idea of having to play games at recess or gym class was too much for me. I made a big distinction between intellect and athletics. I always felt that I was a head person and not a body person. Most of the boys in the little Wisconsin town where I grew up were very jock-y. Since I wasn't that, I kept to myself and read a lot and drew pictures. Fortunately, my family never gave me problems about who I was."

Still, Jim was anxious and unhappy during childhood. He remembers crying in the school bathroom in third grade because some boys had mocked him. After that, he practiced a tougher swagger and spiced up his speech with words like "shit" and "pussy."

Things changed in junior high. For starters, there was no more "recess," so he wasn't forced to play ball games; and for another, he found a new role for himself. The boys and girls started mixing with one another, awkwardly, and Jim served as a go-between. He was handsome, but he didn't go after girls sexually, and thus the girls considered him both appealing and trustworthy. For their part, the boys appreciated having a guy around who was neither a nerd nor a competitor.

Still, the idea that he might appear effeminate was abhorrent to him. Once the other kids started dating, he made sure he always had a girlfriend—Catholic girls who, all the boys knew, would never let anyone past first base.

So it was with great anticipation of ending his long years of inauthenticity that he went public with his homosexuality midway through college. He'd had one affair with a man prior to that time but had kept his feelings secret.

"I'd been active in the ecology movement on campus, so I decided that a good way to come out would be through politics. I joined a gay rights group, and of course those were guys who were immersed in gay culture. Most of them at that time were very hard types, and I didn't know what I was getting into. They told me that it would just be a matter of time before I would become a sophisticated S-and-M'er. In a couple of months I would understand why it was correct to be tough and wear leather all the time.

"I tried to make it happen," he laughed, "but there was no way. It took me a few years and a set of barbells to accept that that wasn't me. It's taken even longer to accept the fact that gay men expect one another to dress in tight shirts and tight pants that emphasize their asses and their chests and their dicks. I mean, I've gotten comfortable dressing like that to be camp at a party, but I wouldn't dress that way to walk around Castro Street or go to work."

Instead, Jim dresses unusually "straight," even preppy. For our meeting he wore a cotton V-neck sweater and loose-fitting slacks, neither of which threw into relief any part of his body. On the other hand, Jim hasn't exactly stayed undeveloped. Partly as a result of the AIDS epidemic, the strong-and-healthy look is very much the order of the day where Jim lives. In addition to the barbells he bought in college, he owns a small trampoline and a sit-up board, and while he doesn't relish the thirty minutes every morning he spends exercising, he admitted it's made him happier with himself and has kept his partner interested.

Biceps Make the Man

Gay men are by no means the only ones to have experienced conflicts over a lack of muscles. A national survey of 62,000 readers of *Psychology Today* found that a man's self-esteem correlates directly with having a muscular upper body. And in experiments in which male college students are given weight training, as the men grow stronger they become more outgoing and their degree of satisfaction with themselves increases.

Yet there is great variation in how men cope with the physical ideals placed upon them. Some men devote most of their lives to building up their bodies, while others scarcely exercise at all. Generally, a man's choice of one of these options or the other, or something in between, depends on what other people made of his body earlier in his life.

Those who suffer as adults are men who somehow never got into athletics while growing up but always felt parental and community pressure to do so. "I make a great pretense of being happy with these arms," said Larry, the thirty-six-year-old owner of an advertising agency in Atlanta, as he demonstrated how thin his left arm is by cupping the thumb and middle finger of his right hand almost completely around his upper arm. "I kid my friends who work out. I tell them, 'Biceps are just ugly bulges.' But the truth is, I'm not happy with my body.

"There's an event that sticks in my mind," he continued. "Nineteen seventy-two. We'd just graduated from Oberlin, and about a dozen of us took over the summer house of somebody's parents on a private lake in upstate New York for a week. One afternoon they all decided to go skinny-dipping. I begged off at first. I don't take off my clothes even in front of people I'm totally comfortable with. I make love in the dark when I have a choice in the matter. But those were the days of free sex and do-your-own-thing, and it wasn't considered cool to be hung up about nudity. They hassled me until I finally stripped and jumped in the water.

"It was about the worst experience of my life. First off, the other guys all had better bodies than I did. My stomach stuck out, even then, and I had no shoulders or chest, and of course, no biceps." Larry laughed nervously.

"The thing that really did me in was when an ex-girlfriend of mine swam by with the man she was living with at the time and made a comment about how I won the funniest-shape-of-the-day award. It was as if someone had run me over with a Mack truck. I felt embarrassed and betrayed."

Fifteen years later, and Larry still has a puny body he's ashamed about. Now not only doesn't he go skinny-dipping, he doesn't even go swimming. But he's not unattractive. In fact, he has a pleasant face and a full head of wavy black hair. One could easily imagine that if he stood up straight and added an inch or two of muscle in strategic spots, he'd look great in the stylish clothes he wears.

Why doesn't Larry simply work out a few hours a week so that he can feel decent about himself physically? He was unable to answer that question directly. The answer came out, nonetheless, at another point in our discussion, when he described his parents and the nature of his relationship with them as a child. His mother used to criticize him for not being the son she'd imagined having, but anytime he showed some independence or virility, she was unsatisfied with his performance.

As he described it: "Either you played ball with the other kids or you weren't a Real Man. My mother's brother Don was a Real Man. He'd been a guard for the basketball team. He's very tall and very fast. He ran a marathon this summer to celebrate his sixtieth birthday. My mother was very pretty in high school, very 'popular,' and she wanted me to be the same. If I'd had a sister, or even if there'd been another boy for her to lay it on, maybe I wouldn't have felt so pressured. I think I wimped out of sports just to spite her constant chirping about how I ought to be more like my Uncle Don."

At the same time, Larry's father, a man who was already distant, became even more so on the few occasions when Larry shelved his stamp collection and put on a baseball mitt. And at an early age Larry noticed that his mother was affectionate with his father only when his father was sick, a handy trick Larry came to deploy himself.

It's clear that in the crevices of Larry's adult mind there lives the belief that he cannot be fit without losing the attention of those he depends on. In his experience, to be fit was to yield to the wishes of an overpowering mother, whereas to be weak was to gain attention from the most important man in his world. This is just the opposite of many boys, who seduce Mom and buddy up to Dad by playing sports.

Larry survived high school thanks to an extracurricular activity that allowed him to relate to his father and to other males. He took up photography, a longstanding hobby of his father's. Working for the school yearbook, he was assigned to photograph football and basketball games. He became friends with other staffers, and in his senior year he was appointed editor. "The yearbook room was my safe zone, the camera was my weapon," he said.

At artsy Oberlin College in the late sixties, his camera attracted the attention of desirable women. It wasn't until that episode at the lake that he started to pay a price again for his physique.

He was safe in high school and college in part because the culture had changed. Some decades are better than others for men like Larry; fashions in brawn wax and wane. During certain periods, American body trends reward less muscular men. Historians have documented several such periods, including the years just after the Civil War, and the 1960s. At other times, including the early years of this century and the seventies and eighties, American men have been required to be overtly strong in order to be received as attractive and healthy.

Physical fashions for men reflect national political trends. During the Vietnam War, men's bodies took on special significance. In fact, the war was *over* bodies. Each side claimed victory less on the basis of territory taken than on "body counts." Those who opposed the war actively deployed their bodies in the service of opposition. Some were beaten up in protests in Chicago, and many more recast their bodies into symbols of defiance—wearing long hair, beards, and odd clothes that distinguished them from Marines. Suddenly, men who'd enjoyed athletics in high school were viewed as no sexier than their comrades who'd earlier been teased for throwing a ball like a girl. Muscles didn't necessarily contribute much to an antiwar image.

After the war, the oppositional look largely disappeared. The current ideal American male body stands as a symbol of reunification. *We have the same basic values*, the post-Vietnam body proclaims, and these are manifest in the trim, strong figure we admire in our men (and, to a limited degree, in our women). *Our goals are identical*, the post-Vietnam body reassures: liberal or conservative, black or white, we just want to be secure and prosperous and in charge of our own destiny. The American body politic, once torn asunder, is mended.

"Muscles have come to *mean* something again: an obsession with the beauty of health and a growing impatience with having sand kicked in our face have combined to give back to muscles a national symbolic credibility," Charles Gaines observed in *Esquire*.

But let's return to Larry, who hasn't fared at all well during the age of brawn. Just after college he married a graphic artist, who helped him set up his ad agency until she grew bored and status-hungry and went back to school for an MBA. A few years ago, she left Larry for someone she met at the health club where she has a corporate membership.

According to Larry, it's hard to have much success on the singles scene when you're out of shape. An analysis I conducted of "personals" ads in ten newspapers and magazines from across the U.S. and from London bears him out. Words like "athletic" and "well-built" appear in a majority of the men's descriptions of themselves and women's descriptions of their desired mates.

Women who advertise in these publications are primarily upper-middle-class, well educated, and looking for men of the same stripe. Since they've broken out of traditional roles themselves to some extent, they might be expected to be more receptive to less traditionally masculine men. In fact, they often prefer to have rather macho men around, perhaps to offset their fears that they may not be sufficiently feminine.

Christine, for instance, the corporate vice-president, made it very clear she has no interest in "pale hairless guys who make great pasta," whom she calls "newts." She dates tall, well-built, handsome fellows. "I don't need to be taken care of," she said, "and I can forge my own way and make a lot of money. I'd sort of like the feminine side of me reinforced by being with a man who is more male than I am. Dealing with a man who has a real female side is unsettling."

Politically left-leaning women can also be suspicious of men with sunken chests. One woman I interviewed, who has refused to wear makeup her entire adult life on the grounds that the cosmetics industry is a capitalist plot to enslave women, said sternly about men: "The obligation to be beautiful is oppressive, the obligation to be strong is empowering. A woman who refuses to 'fix her face' is simply rejecting patriarchal oppression, but a man who refuses to build up his body isn't making any kind of statement at all, except that he's lazy."

Fear Makes the Biceps

Given their poor reception in the outside world, men who are physically weak understandably experience low self-confidence. What's surprising is that their mirror opposites—the hunks and superjocks—often suffer from the same problem.

Perhaps the single greatest force that keeps men working out is insecurity. This is evident in those who exercise chiefly because they're afraid of heart disease. But almost all avid male exercisers are engaged in a passionate battle with their own sense of vulnerability. Herein lies an important distinction between men and women. For both, the key motivation to exercise is improved self-esteem, but the genders differ on what they believe produces these benefits. When surveyed as to why they exercise, women talk about accomplishment, beauty, affiliation with others; men say they're motivated by the chance to pit themselves against nature or other men and to confront physical danger. In other words, men seek to prove to themselves and others that they can survive, that they're winners.

The harder a man exercises, the more he may be trying to overcome his feelings of inadequacy or helplessness. Most bodybuilders in a study conducted in southern California were found to have been stutterers, dyslexics, thin, fat, short, near-sighted, or otherwise unacceptable to their parents when they were children. The author of the study, sociologist Alan Klein, proposes that bodybuilding sometimes serves as a kind of "therapeutic narcissism." Through it, those who feel insecure are offered a way to devote their full attention to making themselves big, strong, and commanding of attention.

I developed a vivid appreciation for the sweat-for-salvation aspect of male fitness when I visited a place where a high concentration of America's best-developed men live—a maximum security prison. There I met a man named Nathan, who is famous in several California prisons as an advocate for strengthening and perfecting the body while in jail.

Attractive and well-groomed, Nathan wore a short-sleeved yellow Lacoste shirt along with his starched gray prison pants and gave off a scent of expensive cologne. His closely cropped beard was cut precisely to complement his square features and his short curly black hair, which had obviously been styled by a talented barber. (He had an arrangement, he explained, with an inmate who had worked in a Hollywood hair salon prior to his conviction on drug charges.) Nathan's huge arms, covered with blue tattoos of eagles and naked women, offered a strange counterpoint to his fastidious grooming.

In a small room off the main visiting area, I asked Nathan

to describe the different types of men who build up their bodies in jail.

"You got the superheavyweights, over six feet and massively built," he began, over the constant hum of prisoners yelling from the cell blocks in the adjoining buildings. "Then you got the real short guys. The tall ones are usually pretty smart— they don't have a college education, but they have common sense. They don't want to be overtly aggressive, they just want to keep people away and do their time. The short ones usually are abrasive. They're looking for trouble. They've got that Napoleon complex. Then you've got the guys that want to box. They go through a very rigorous boxing discipline. They run, they practice all the boxing techniques, jump rope, hit the heavy bag."

Nathan estimated that two-thirds of the inmates at the prisons he's been in are seriously involved in exercise of some type. "When you come to jail you have a lot of time on your hands," was his explanation at first. But as he talked on about prison life, and his own biography, a more complex picture emerged.

"When you're in the streets," he said, "you have a lot of time, but it's not structured into roll calls and meals, so it seems to go very fast. In here you're really conscious of time, and one of the pastimes that you can see some results from is lifting weights. You see people around you and you say, 'Wow, that looks nice, that guy has a nice build.' The way he carries himself, the way he walks, the way people respect him. And when you get big it gives you an artificial sense of security."

In what way is it artificial? I asked, feeling oddly comfortable after only ten or fifteen minutes with this Herculean man whom I knew to have been convicted of murder.

"It's artificial because you have to defeat the fear within

your heart," he answered. "How big or small you are doesn't have anything to do with it. I had nineteen-and-a-half-inch arms at one point, but I couldn't pacify the fear in my heart, and people could see that."

His personal fear, he went on to explain, is that he'll spend his entire life in jail. He was first locked up, in a mental hospital, when he was five and a half years old. On that occasion he'd been playing with matches; he set fire to a sheet and his family's apartment went up in flames. His father, who was confined to bed for a back injury, died in the blaze. "My mother wanted to love me," he said with practiced dispassion, having relayed the story many times, "but she couldn't because she blamed me for the death of my father."

Nathan was cast out by his mother and didn't fare much better in his neighborhood. His light-brown skin and his ethnic background marked him for trouble from the time he was a young child. His father had come from the Cape Verde Islands and his mother from Brazil. In the barrio where Nathan grew up, "people had names like Carlos and José, and here I was Nathan. Kids used to think I was white because I'm so light, and there I was being acculturated into the Chicano culture, yet I couldn't identify with them physically. Everywhere I went it was understood I wasn't one of them."

Most of his formative years were spent in juvenile detention facilities. He was angry and confused, and he struck out with acts of violence ranging from schoolyard fights to armed robbery.

During one of his longer stays on the outside, at age twelve, Nathan shot heroin. He continued off and on until he hit forty, when he took up yoga and physical fitness in prison. The inmate who taught the yoga course espoused the view that drugs are poison and drug users pathetic creatures.

After his release on parole, Nathan became a community

crusader against drugs, combing the streets for strung-out kids he could Pied Piper into his martial arts classes, which he conducted free at a recreation center in Watts. The more respect Nathan got in the neighborhood from his physical abilities, the more grandiose he grew, and within a few months he was preaching about "eradicating drugs from the face of the earth."

One afternoon, when an adolescent follower arrived with the news that another was in a coma from an overdose, Nathan went looking for the drug dealer who had sold him the stuff. He beat the man up badly and left him bleeding in an alley. The man died a few hours later, and Nathan was sentenced to twenty-five years to life for murder.

"Once I was back in the joint," Nathan remembered, "I borrowed some law books. I knew I'd spend the rest of my life in the joint unless I could find a way to get around this sentence. But it was hard to concentrate because I was on an open block. You have TVs and radios on full-blast and people yelling twenty-four hours a day, seven days a week. I had to be able to pull myself inward and digest this material, to think of an approach to use at my defense."

Nathan initiated a daily regimen of physical development and purification which he still continues. He says it gave him the willpower to study the law and to argue successfully before the parole board in 1985 that his sentence should be reduced to six to twelve years.

Today, Nathan's routine goes something like this. He rises at 5:00 A.M., when the cell block is still reasonably quiet. Without making enough noise to wake anyone, he repeats twelve times each a series of special exercises that combine calisthenics and yoga. In describing these to me, he left his chair and demonstrated. Assuming a squatting position, he took a very deep breath that expanded his chest muscles to their fullest; he held this for a few seconds, then gracefully raised himself upward to a full standing position, from which he bent forward while slowly exhaling, until his palms touched the floor.

On his way back up, Nathan caught a glimpse of the concerned expression on the face of the guard outside our cubicle and sat down again. He continued his description: "As you see, I don't look anything like the yogis. If you see them in a book, they look like they're malnourished, whereas my body is well-developed everywhere. Yet I can put my elbows on the ground from a standing position with my knees locked. I'm superflexible. In America you want to have a good, healthy, rich image, not malnourished. My concept is that you can maintain that look and at the same time have flexibility."

After his predawn exercises, which take an hour, Nathan eats breakfast in his cell. He refuses to eat in the mess hall he said, because the food isn't healthy. Instead, with money or cigarettes earned from advising other inmates on legal matters, he orders health foods through the prison commissary. Friends who work in the mess hall also bring him milk, fish, and vegetables a few times a week.

After lineup, he attends a class offered in the prison by a local university. At the juvenile detention facilities where Nathan grew up, schooling was provided only a few hours a day; the rest of the time Nathan hauled coal and cut grass. His formal education was poor at best, and so it's no minor accomplishment that at the time I met him, he was about to graduate from college as valedictorian of his prison class of twenty-eight.

He credited his educational achievement to his bodily discipline. "It really opened my mind and gave me a sense of direction and a focus for my energies," he said. Classes in the prison are offered in the mornings and early evenings. In between, nonstop from 1:00 until 5:00 P.M., Nathan can be found in the jailyard working out with a group of inmates who have taken up his fitness system. They run a few laps to limber up, then move into the same sorts of exercises Nathan performs alone in his cell.

Life in a hot, violent, noisy prison is hell for anyone. Still, Nathan has a busy and secure life behind bars. "This has become like a womb for me, where I can function and be successful," he let drop at one point in the interview. Each time he's been released, he's come unhinged within a few months. As an adolescent, "just having been in prison was a status symbol. When I went home they had a party and everybody said, 'There goes a sure-enough bad dude.' I felt great. Then, two days later, I had to prove myself all over, and I'd get into trouble again." As an adult who has spent so much time behind bars, he says he can't maintain routines when he's on the outside. "I overindulge. I stay out every night dancing and having sex, trying to make up for lost time. I become totally fatigued, and then I feel bad because I'm not keeping my mind and body as sharp as I know I'm supposed to. I get paranoid and confused on the street."

It may be ironic, and it's surely unfortunate, but Nathan feels more at home in jail than he does on the outside. Behind bars he can maintain some measure of self-respect and control, thanks to his fitness regimen. His exercise program has given his life order and predictability and is a source of personal pride.

Calming the Storm

Among the law-abiding men I interviewed who exercise obsessively for periods of weeks or months, the same basic motivations apply. They discipline themselves through fitness in order to stave off the impending chaos they confront in their daily lives.

A case in point is Roger, a forty-two-year-old Chicago lawyer. While growing up, he played softball in the neighborhood; in college he did nothing beyond a morning wakeup routine of push-ups and sit-ups; and in law school he "hardly had time to eat." Although Roger was in the top quarter of his University of Chicago Law School class, he was terrified he'd fail the bar exam, as his older brother had. He countered his fears by bingeing on exercise. "I've played racquetball exactly twenty-two times in my life, all within the last three weeks before my bar exam," he reported. "Racquetball made me feel better. When I couldn't study, it loosened me up. I didn't play particularly *well*, but nobody played *harder*. I broke several racquets and messed up my arm pretty badly a couple of times."

That was the first of three times in the past fifteen years that Roger has gone on and off the exercise wagon. From the day he received notification that he'd passed the bar, about the only exercise Roger got was lifting heavy law books in the back offices of a large firm—until, that is, his second exercise blitz, which began a couple of years out of law school when he found himself unable to sleep the nights before he was to do battle with another lawyer in court. He'd awaken at four in the morning, his jaws clenched and his stomach knotted. In the middle of an argument in the courtroom the next day, his mouth would go dry and he'd lose his train of thought. Once a judge asked him to approach the bench and in a peevish voice advised Roger to request that his firm send him to a public speaking course.

Instead, Roger took up running. On his way back from court, he bought an expensive pair of Etonics, and each morning thereafter he ran three or four miles around Grant Park before going to his office in the Loop. "My father had had a heart attack at a young age," he said. "With all the reports coming out at that time about the benefits of running, I decided to join in." Before long he was doing eight miles a day, then ten and twelve. Some days he didn't feel like running but would run anyway: "A mile into the run I would get that feeling of relaxation, of going on forever, a real kind of power." And he was able to sleep at night and to present a strong case in court.

By the end of his third year out of law school Roger was, in his own words, "a damned good trial attorney." Within a year, his confidence firmly in place, the running fell off to a few miles every other day, then diminished to nothing. He blames the long hours at work required to make partner.

Except for a taste of golf, which he found boring, Roger again went without exercise until a year before our interview. The event that precipitated this, his third exercise spree, was sudden rejection by a long-term girlfriend. As an antidote to the pain and vulnerability he felt, Roger hired a private exercise trainer known in Chicago executive circles to be unsparing, even sadistic, in his drive to get his clients back in shape. For ten months prior to our meeting, the trainer had greeted Roger every Monday, Wednesday, and Thursday at 6:00 P.M. upon Roger's arrival home from work, and Saturdays at 4:00 P.M. For a grueling hour he orchestrated Roger's workout in the gym they set up in the spare bedroom of Roger's Lake Shore Drive apartment. "I never would have thought it possible that I could be in such great physical condition," Roger claimed. "I feel great and I look great."

Nevertheless, when I met him he was already showing signs that the end of this current exercise cycle was in sight. He told me of plans to decrease the number of training sessions each week; and although he "swore off women" after his disappointment the year before, a few weeks prior to our interview he began sleeping regularly with a thirty-two-year-old physician.

Men who exercise for purposes of deliverance (like Roger and Nathan), as well as those who abstain from exercise (like Larry), differ in an important regard from men who are at neither of those extremes. Exercise is not a highly charged activity for those who pursue it in a more moderate way. They don't attach magical significance to lifting heavy objects or hitting balls.

A hallmark of a sane exercise program is that it is integrated into a person's daily life. It's just something a man does, like eating lunch or getting a haircut. He goes to the Y or health club regularly to play basketball or handball or pump a little iron with his buddies. And although his sports activities may take up a fair amount of leisure time, he forgoes them if a family or business emergency takes precedence.

At times—when he's angry with someone at work, for instance—he may play rough and injure himself, but the displacement of frustration is not what his athleticism is about. The role of athletics in his life is much more basic than that. Typically he has been involved with sports since childhood. He loves to reminisce about a particular game from his youth, or the day his dad installed a hoop on the garage at the end of the driveway when he was five or six. He has a vivid memory of his father placing the massive basketball in his arms and lifting him up so he could sink it through the basket. In the twenty or thirty or forty years since, there's never been a period when he hasn't played some kind of sport; in high school he may even have made it onto a team. And he devotedly follows college and pro teams on TV.

Lifelong jocks are living evidence for a current view of human development called, appropriately enough, "continuity theory." It holds that our interests during adulthood are usually extensions of what we enjoyed as children. Continuity theory disputes the myth perpetrated by sports magazines and health clubs—that a devoted couch potato can, with a bit of willpower, transform himself at age forty into a championship marathon runner or ball player. Men who try athletics for the first time during adulthood seldom succeed; like Roger, they don't stick with any activity very long.

Several studies show that men who engage in exercise on a regular basis as adults also did so during their childhood or adolescence. One of the best predictors of whether a man will be athletic in midlife (and later) is whether his father participated in sports and brought him up to do so as well.

Men who've grown up athletic are the great beneficiaries of the American male role. When social scientists track down high school athletes ten or more years after graduation, they find them holding better-paying, higher-status jobs than their classmates from similar socioeconomic backgrounds. Their body image and self-esteem are greater, too.

Who can say whether these positive outcomes are the result of their athleticism or are coincidental with it? Whichever it may be, other men envy these men their comfort with their masculinity and their physiques, and women wish it were as easy for them to stay pretty as it is for these men to stay handsome. How unfair, I've heard women complain, that such men need merely continue to play the games of their youth, while to maintain their beauty women must spend hours in beauty salons having perms, facials, manicures, and pedicures; must starve themselves on diets; must wear uncomfortable shoes . . . and on top of all that, exercise whether they enjoy it or not.

Although some men do fit that picture, they're a small minority. Most men, even if they've kept themselves reasonably fit, are privately insecure about their looks and more vain than others imagine them to be.

SEVEN

Power and Vanity

"The Sexiest Man of the Year." Last year, according to *People* magazine, he was Harry Hamlin, the star of *LA Law*. Not an easy subject to photograph, apparently. "With the Joan Collins cover," said Harry Benson, veteran *People* photographer, "I shot two Polaroids first to test lighting and looks, and Joan said, 'Fine.' Hamlin was perfectly pleasant, but it still took fourteen Polaroids before he was reasonably satisfied."

I've heard similar comments about male models. One of New York's leading stylists told me: "A lot of times I want to smash the mirror on these guys. I say, 'You look fine, believe me. If you looked awful we wouldn't want to photograph you.' If a single hair is out of place, they go nuts. Women are much easier to work with."

Are actors and models, because they're vain, different from other men? I think not. Like the quaint notion that the wealthy are unhappy, the idea that men don't worry about their looks is a myth.

In a national survey conducted in 1985, roughly the same number of men and women (34 percent compared to 38 percent) said they were dissatisfied with their appearance. When

I surveyed students at Syracuse University, only 10 percent of the men said they were "not concerned" about their appearance, and 42 percent indicated they were "very concerned." A study at New Mexico State University found that fully 70 percent of the men questioned were unhappy with their physiques.

At the other end of the age spectrum, about 96 percent of both the men and the women who filled out questionnaires at senior citizen programs in Philadelphia indicated a continuing concern about their appearance. Both sexes said they felt good when they looked good, and roughly equal numbers expressed a wish to change some part of their body in order to better resemble society's ideal.

Traditionally, a man's concern with his face or his dress has been something he's kept to himself. He might admit it confidentially on a questionnaire, or he might buy a few "grooming aids" on his lunch hour, but to the world at large he shows no vanity. To do otherwise would be to appear unconfident, effeminate, or just plain conceited.

But lately, at least in some quarters, male vanity is growing more visible. In "personals" ads in the mid-1970s, far more women than men advertised themselves as attractive, but this pattern no longer holds. In the mid-1980s, I found in my own study of the ads that the emphasis on looks is only somewhat greater for women than for men—51 percent of the men and 63 percent of the women claimed they were attractive. At the same time, 61 percent of the men requested photos from the women they hoped would eventually write, and fully 56 percent of the women asked for such visual aids. These numbers suggest that men know that looking good matters, and that women select their dates on the basis of looks as well as other characteristics.

In suggesting all this, I'm not making light of how our

patriarchal society forces women to trade on their appearance in order to please men. My point is simply that men also use their appearance, though to signal success more often than beauty. Either type of display is carried out by way of the body. The CEO of a major corporation told me, "Guys who make it in business these days work their butts off for the company, more than ever before. But they *look* as if they spend their mornings shopping for suits, their lunches working out, and their afternoons at the hair stylist's.

"A guy who looks rusty," he said, "is going to be bumped off by an ambitious younger guy who looks sharp."

Throughout their lives, men are given signals that their looks do matter, although they shouldn't act vain. Better-looking men are typically more self-accepting and confident, and they feel more in control of their lives than less attractive men, studies find. They're also valuable commodities for women. Everyone knows about "the Onassis effect"—homely men whose wealth attracts beautiful women. But how about "the Collins effect"—Joan, that is, who enhances her looks by keeping handsome men beside her in public? A woman is judged more favorably when in the company of an attractive man.

In the work world, good looks actually appear to be more broadly valuable for men than for women. Handsome men have been found to have an advantage in landing both clerical and managerial jobs, while attractiveness improves women's chances in snaring only clerical positions. Beauty can actually be a liability for women seeking executive jobs.

Yet despite the importance of handsomeness for men's self-esteem and material success, establishing themselves as attractive may be more difficult for men than for women. First off, men are not as readily viewed as attractive. On the average, women's faces are judged more attractive than men's. And a study of married couples found that husbands typically consider their wives better-looking than other people do, while their wives rarely return the favor.

Also, men are not as well equipped to enhance their looks. "Women in our culture probably learn early to take stock of the strengths and weaknesses of their appearance and to take whatever steps are possible to bring their appearance up to established standards," write social psychologists Cynthia Rand of Johns Hopkins and Judith Hall of Harvard. "Men are much more restricted than women in what they can do to improve their attractiveness via hair and clothing, and indeed may feel that excessive attention to such matters is unmasculine."

Mundane Vain

Even though men spend a billion dollars annually on toiletries, the primary way in which they enhance their physical appeal is not by decorating their bodies but by firming them up. The difference between male and female attractiveness is the difference between the words *handsome* and *beautiful*. *Handsome*—the adjective more often applied to men—connotes, according to the dictionary, an "imposing appearance suggestive of health and strength." *Beautiful*—a word used primarily to describe women—means "capable of delighting the senses or mind."

A man makes himself better-looking by appearing more powerful, a woman by adorning herself. There are certainly men who spend great sums of cash and energy on their hair, their skin, and their clothes. But to understand the nature of ordinary male vanity, such men are the wrong ones to turn to. Better to talk with someone like Edward, a fifty-three-year-old prison administrator in Maryland who wears

J. C. Penney suits and has an eight-dollar barbershop haircut.

Edward is not *perceived* as someone who pays much attention to his appearance. In fact, that's why I interviewed him. A friend had insisted that my pool of interviewees must be skewed if I thought men cared about their looks. Out in middle-aged middle America they don't, she said; and she nominated Edward as someone I should meet.

Initially I was struck by Edward's rigid, official demeanor. But after I'd spent an hour with him, and he trusted me not to repeat his comments to his associates, he spoke openly when I asked how he feels about his body. Pointing to the thinning hair atop his head, he said, "People who haven't seen me in a long time will say, 'You're getting gray,' and then they'll hasten to add, 'but it looks very distinguished.' They're conscious of my feelings, I guess. I think about dyeing it, but then I'd have to justify that to everybody.

"But I can't be worrying about my hair. The thing I need to concentrate on in the near future is this blubber belly," he went on, poking an index finger an inch or two into his middle. "Top priority is to lose some in here. I'll feel real good about myself when I take the weight off."

Edward told me in considerable detail about his shame over the three inches that hang over his belt, and how he bought a brand of pants specifically because their ads promised to fight "roll-over." ("All it was," he complained, "was a stiff waistband that bent the first time you sat down.") He has sworn off beer completely, and he religiously performs fifty sit-ups every morning. He even bought one of those $39.99 tummy busters advertised on a cable sports channel.

Later in the interview, Edward detailed for me the care he takes in ensuring that his shirts are properly starched and his pants creased precisely down the center of each leg.

Edward's attention to the details of his appearance point up the true nature of garden-variety male vanity. The ideals of attractiveness for most men in America are military ideals. The goal is to appear disciplined and orderly—shipshape. At that Edward succeeds. Despite his gut, at just six feet, he's as solid and imposing as the huge desk that separated us for the three hours we were together. For the entire time he sat perfectly straight, his rock-hard shoulders pressed firmly into the back of his chair.

He told me: "In any line of work you're a model. You're a role model for the people who work under you, and you're a role model for the people you serve, whether they're customers, in a business, or the inmates in this facility. Especially for the young inmates, you're a model for how they feel about themselves. You have to give them the message that how you take care of your body, the way you dress and carry yourself, says a great deal about your self-esteem.

"I've always respected people who appear to take care of themselves, and I'm a stickler about my staff's appearance. That doesn't mean I'm anti-beard or against long hair or opposed to any particular style of dress. But they had better look clean and neat.

"The public has a right," Edward declared, "to have its public servants look respectable. When I see someone who's messy, I get angry. I'm paying his salary, and I expect more. Neatness is only a step away from cleanliness in my book."

Anthropologist Mary Douglas has shown in her research that throughout the world people believe the equation *order = cleanliness = safety*, as well as the flip side, *disorder = dirt = danger*. The elaborate rituals people undertake to look tidy—from ironing their clothes to centering the paint on their foreheads—are actually efforts to keep social danger and chaos at bay. If a man in a responsible position appears put-together, he signifies that his workplace is stable, that it's not about to fall apart financially or morally.

In almost every setting, well-groomed employees are assets,

but in prison work, impression management is crucial. Control, order, and respect—these are the products the organization sells to inmates and to a skeptical public. A man with Edward's values and appearance is ideal for the job. He's the very embodiment of all three traits—traits he's inculcated in himself since adolescence, in large measure as a response to what he experienced as a child.

If, as we saw earlier, women live out their mothers' dreams through their own bodies, so do men their fathers'. Edward's father, a Ukrainian immigrant with big plans for wealth and respectability in America, scrimped for a decade in the tenements of lower Manhattan before saving enough to move his wife and two young sons to a town in northern New Jersey. There he set up a roofing and carpentry business, only to find he couldn't compete with the Italians who had family capital and connections. The more he was forced to take low-paying jobs with local contractors just to get by, the more grandiose his talk became.

Before long, Edward's father could sustain his dreams only with the help of alcohol. He played the role of big spender at the sole place he could get away with it, the local bar where he bought rounds for everybody. Edward has vivid memories of nights when his mother sent him and his brother to the bar to fetch their father. "I felt so *dirty*," he said sternly, "and small."

He was a short kid anyway and got knocked around on the way home from school by black kids from across the tracks and white kids from his own neighborhood.

During adolescence, Edward's brother aligned himself with their father, assisting him on jobs when he worked and defending his drinking at home. (Today, at fifty-five, he's still walking in his father's footsteps, working for a roofing concern and fighting his own drinking problem.) Edward, meanwhile, resolved to better himself. Like many eager kids from poor families, he turned to his body as a vehicle for escape from his family's despair. He enrolled for free boxing lessons offered at a community center a mile walk from his house. Every afternoon, Monday through Saturday, he could be found punching the bag or jogging.

About half a year into his training, one of the neighborhood bullies challenged him to a fight. "They put us in a ring together," Edward recalled proudly, "and I practically knocked him out. I was fired up. I'd been wanting to bust some guy open for years. And I became an instant hero in the community. By the time I graduated from high school I'd had eighteen amateur fights and lost only two."

Boxing changed Edward dramatically. He began as a light-weight, yet by age eighteen he'd become a 190-pound heavy-weight. After high school he joined the Army, which sent him to Germany, where he gained attention quickly for his boxing abilities.

During his second year there, he was pitted against a black Marine. "The white guys were all betting on me, and the black guys on him," Edward said. "I knocked him out very quickly. I don't have the foggiest idea where my power came from, but I did more than just knock him out. In the process, I fractured his jaw in a number of places. He was unconscious for ten minutes.

"The white enlisted men were cheering for me, but I was watching the medics work on him. They finally got him to his feet, and he didn't know what was happening. He thought he'd won the fight. He started jumping up and down with his hands in the air, and then he fell again. They had to call a stretcher and carry him out.

"I'd never seen this man before in my life. I had nothing against him. I found out two days later that he'd had to have his cheekbone wired and plastic surgery done—it would be with him for the rest of his life. I told the coach that I would never fight again, and he said, 'You'll get over it, kid.' He saw me as one hell of a prospect for all-Army and possibly the Olympics. I had very fast hands."

As Edward thought about the star he might have been, a wistful look came to his face. He promptly erased it, saying: "I had no choice, I had to get out. Maybe I had the physical tools a boxer needs, maybe not, but I didn't like what it was turning me into. My whole body, my whole mind were bent on knocking this guy out. I had greater ambitions for myself than that."

The story of the years since his discharge is all about a personal struggle to build a more tolerable persona. Lacking a college education, and looking and speaking like a tough immigrant boy, he couldn't land the sorts of jobs he wanted. Only after being rejected for dozens of sales and assistant manager positions did he take a job as a prison guard.

At that first prison, his reputation had preceded him. Some of the inmates recognized him from the time his picture had appeared in *Ring* magazine. At first he didn't discourage them from viewing him in that context. "I was twenty-two," he said, "and in those days the inmate population was older than it is now. They would have made ground beef out of me if I didn't have their respect."

Toughness is a valuable attribute for someone in prison work, and one that Edward has maintained throughout his career. He's generally considered "a tough boss." On the other hand, what carried him out of the cell block and into the superintendent's office was not his fighting ability. He enrolled for every staff development course the system offered and consciously emulated the manner of dress and speaking of his supervisors. He even shifted his sports interests, from working-class favorites like boxing and drag racing to more refined, middle-class games such as basketball and baseball.

Asked recently by a local reporter for the secret of his success, Edward listed, in proper macho style, his ability to make difficult decisions and his personal drive. But late in his interview with me, he phrased matters in different terms. "When somebody on my staff does something special, I put him out front and recognize him. We had one guy this year who used his CPR training to save another guard's life. I had a special ceremony and called in the press. I make sure this place isn't just another warehouse for felons. It's a question of personal dignity, the way I see it. You have to remember, I'm still fighting off that memory of my dad. When he had his shop, it looked like a hurricane had swept through it. Supplies and papers were scattered all over the place, and he looked a shambles himself."

Edward reached into the side drawer of his desk at that point, and removed a Norelco rechargeable shaver. "I also keep a fresh white shirt hanging in that closet over there," he said, pointing across the room with the shaver, "still in the plastic bag from the cleaners. I've got a heavy beard, and I tend to perspire more than a lot of people. If I have a politician or a reporter or somebody coming out to meet me in the afternoon, I don't want to have a Nixon shadow or be wearing a shirt that has stains on it. Nobody trusts a man who looks as if he walked out of some old gangster movie."

"The Worst Feeling a Person Can Have"

When we think of people evoking a response on the basis of how they look, and trying to adjust their image in light of others' expectations, we think most often of women. But the struggle to be seen differently, to make more of yourself

143

than your face or your roots easily allow, is a theme in the lives of men as well.

Those whose bodies fit a cultural stereotype are likely to find that their lives are governed in large part by the assumptions and expectations of others. The course of their relationships, the job choices they make, the nature of their self-doubts, all are propelled by the powerful preconceptions people hold about them and their "type."

I've talked with fat men who have spent a lifetime proving to others that they're just as intelligent and energetic as their trim colleagues who get promoted at work. And I've talked with short men about the difficulties they face in demonstrating their sexworthiness to women. But not until I spent a couple of afternoons listening to a man named Charlie, a ranking official with a New York State social services agency in Albany, did I really appreciate what it's like to be at the mercy of others' presumptions based on one's body type.

A strapping six feet, three inches tall, 230 pounds, with muscular arms and legs and a dark complexion, Charlie has been the focus of many people's fantasies about black men, ranging from schoolteachers who expected him to fail as a child to white women who imagined he'd be brilliant in bed.

Our whole first hour together, he answered my questions mechanically, like a man in his sleep, until I asked him something that uncorked him: What was his most vivid memory from his childhood in the South? Fixing his eyes on the office wall behind me as if it were a movie screen, he said, "One time when I was maybe three, four years old, my mother asked me to wait outside the grocery store while she picked up some odds and ends. Some white men came by and noticed me. They made fun of me, called me names, and it ended up with one of them spitting on me. As my mother came out of the grocery, I was in tears, and she asked me what had happened. I told her, and I could see that kind of hopeless feeling in her eyes—an expression of 'I hate that this happened, but there's nothing I can do about it because that's the way things are.' "

Charlie's big deep voice fell away. He kept talking, but even though I was sitting just three or four feet from him, I could hardly make out his words. "It started working on me mentally, I started wondering how long it would be this way," he said as he turned back toward me. "I used to have this one teacher, a white man named Mr. Williams, back in fifth grade. There was one Caucasian in our class, and every Friday around two o'clock this Mr. Williams would ask her to go to the library, and then for the next hour he would tell the whole class, 'You're nothing but a bunch of niggers, that's all you'll ever be, regardless of how much education you get, you'll always be janitors and garbagemen and housemaids.' At that age I felt it was true. My father and my father's father worked for the sanitation department, and my mother cleaned houses, and her mother had been a maid, too.

"I think I will always have a feeling of inferiority," he continued in a low voice. "Now I'm working in an environment where I am able to go where I want to go and do the things I want to do. But I always have some apprehension. I always feel as though I'm being looked upon not so much as a person who happens to be black, but as a black person, and all the stereotypes connected with that. People have certain expectations in terms of how you should act, how you should talk."

At that I blushed a little because I had been marveling to myself at his flawless "white English."

Charlie, thirty-seven, shifted back to recollections from his childhood. His parents split up when he was twelve, he said, and his mother moved him and his brother to Pennsylvania, where her sister lived. "In seventh grade I had a crush on a white girl in my class. We were talking back and forth, and one day I said something, and she jokingly hit me on the top of my head. But once she had touched me—she'd done it impulsively—she took her hand back and looked at it. Then she wiped it on her skirt."

As he described this event, Charlie wrung his own hands together into a strong grip. "I think that must be the worst feeling a person can have," he said. "Knowing that your body is different, and that it's repulsive to another human being, and not even being able to understand why you're different. You find yourself caught up in trying to prove you're all right, that you're as good as they are, and at the same time watching out that they don't hurt you again."

Thanks to a series of events when he was fourteen and fifteen, Charlie's schoolmates came to take a different view of him. It all started one evening when a man arrived at the door and informed his mother that their car would be repossessed if she didn't keep up her payments. His mother tried to reassure Charlie that there was nothing to worry about, but the following afternoon after school, he hopped on a bus for downtown, lied about his age, and got himself a job packing and loading boxes at a warehouse. He worked every day after school, and full-time that summer. By the start of his freshman year in high school, he was twice the size he'd been a year earlier, and he made it onto the school football team.

The next three years of his life Charlie referred to as his glory days. He played both offensive end and defensive end—positions designed for men who are large and fast and can stay on the field the entire game. He dated the school's black cheerleader, and when the team won a regional tournament his junior year, Charlie became a local hero.

But then, in the space of just two or three seconds one morning the summer before his senior year, Charlie's bright light was extinguished. A careless driver backed a Buick out of a driveway and into Charlie, who was riding his bike to work. Charlie landed on his right knee, which has never been the same. With his body no longer a valuable commodity on the football field, Charlie's hopes for a college scholarship were quashed.

His last year of high school was filled with disappointments, he told me. His white friends from the team drifted away, and his girlfriend took up with another ballplayer.

Around dusk one night, sitting on a hillside in a park, he and a couple of black friends were sharing their frustrations over the desolate futures that awaited them after graduation. "It seemed like the race riots were on the news every night that year," Charlie recalled, "and the civil rights movement was at its peak. My two buddies and I were running down how bad things were for us in that town, and as a race of people.

"One thing led to another, and we decided we were going to burn down this appliance store across the street from the park. We found some old wine bottles and stuffed them full of rags and walked down to the gas station and filled the bottles with twenty-five cents' worth of gas. We lit them and threw them right through the window of the store."

As he relayed this story, Charlie became more animated, his powerful right arm cocked back to illustrate the tossing of the grenade. But then his face fell, long and boyish, and his arm dropped back into his lap. "A passerby saw us, and we were caught that same night. My mother was devastated."

He was sentenced to five years' probation and spent the next couple of years knocking around town in a state of rage and defeat. During that period he married a local girl whom he'd gotten pregnant. They moved into an apartment together, but by the time the baby was six months old, the marriage broke up in a violent brawl.

Charlie exited. He joined his brother, "who lived in some of the finest bars in Rochester, New York," and he ran the streets for a year, taking assorted jobs for a day or a week when he needed money to impress a woman or to pay off a gambling debt, and living off his size.

"When you're new in town," said Charlie, "people kind of feel you out, see what your limits are, and my being big and in good shape really cut down on the frequency of that. I messed up a few guys who were ragging me."

He laughed a loud laugh at the memory of himself as a mean dude and went on to relate some of his adventures from that year on the street: the familiar sex, drugs, and fight scenes. Before long he got to the bottom line—one can't truly feel successful on the streets, because the streets are all about failure, filled as they are with people like his younger brother, men (and women, who carry razor blades in their bras) wishing they had respectable jobs and lives and knowing they never will. Charlie got scared about who he'd become, and gradually he disciplined himself with the help of a karate class offered at the local YMCA.

It was the karate instructor who launched Charlie on his career, by passing his name along to a friend who ran a neighborhood halfway house for juvenile delinquents. Charlie's size worked to his advantage in gaining respect from the boys, and over the course of the next several years he advanced from a night-shift attendant to manager of the house. At the same time, he studied for his college degree at night and, in 1984, was hired for an administrative job with a state agency in Albany.

The job stability that comes with a civil service appointment quieted many of Charlie's insecurities but also created a whole new area in which his looks set him up for frustrating cycles of acceptance and rejection. At the agency, Charlie met a lot of women, most of them white, many of whom were intrigued by the possibility of sleeping with a handsome and successful black man. And so he became something of a sex object.

Female Troubles

To have women ask him out made Charlie feel desirable, but also used. "To other men it sounds flattering, and it was, to begin with," he said, echoing what beautiful women sometimes report, "but after a while you tend to lose part of yourself every time you involve yourself that way. You're only ripped off in the long run, and lonely. It tended to make me more aggressive or hostile in my relationships, to be very honest."

Charlie jumped out of the sexual whirlpool two years ago when, after a string of short-term relationships, one of the women he was seeing got pregnant. In a déjà-vu of his first marriage, he felt duty-bound to marry her. Only this time he also decided to try to make the marriage work. A guiding principle Charlie and his wife share, thanks to their careers in social service agencies, is that professional help is a good thing. So they started going to a marriage counselor once a week even before their wedding.

They've worked through many of the problems that stemmed from her assumptions about black men and his about white women. When I met him, they were working together on another set of image problems: their professional images. As both approached forty, they were trying to revise how they were viewed by the outside world.

To that end, Charlie was taking business courses in the evening and thinking about professionalizing his appearance. "There's a big demand right now for black managers, but outside of government, they don't look a whole lot like me. You have to know how to dress right, and you can't look too scary, if you're black."

In his search for an upmarket image for himself, Charlie has taken out a subscription to a fashion magazine for black men and dropped his YMCA membership to join a fancy health club. Executives jog and play tennis, he figures, and so he has traded in his ratty tennis shoes and gym shorts for Reeboks and a Wimbledon racket.

Whether or not Charlie's revised image ultimately pays off in the work world, his efforts to cultivate it have brought him closer to his wife. She has taken up running as well, as part of her regimen to get back in shape after her pregnancy and before she starts looking for a better-paying line of work. And a few times a week, a neighbor takes care of the baby while Charlie and his wife work on their backhand together.

"She's a good tennis teacher," he said, "but I'm not the easiest pupil in the world. Tennis is a far cry from football, and it's going to be some time before I'm any good. I'm very self-conscious on the tennis court. It's another one of those situations where I feel that other people are looking at me and judging me.

"But she's very understanding, and in return I'm trying to be accepting of all her different diets. This past week about the only things we ate were skinned chicken and raw vegetables. I thought I was going to choke. You're talking to a man who grew up on pork. I've been known to eat *two* desserts with my dinner. But I was good. I never once complained, and I made sure I complimented her on the weight she lost."

According to Charlie, he and his wife have managed to take their worries about their bodies and convert them into points of contact. Each is helping the other to change, and they are adapting to one another's concerns.

Husbands and wives are not always so flexible. Many of us, when we feel fat or unattractive, *distance* ourselves from our partner. We blame him or her for our troubles in order to displace feelings of personal inadequacy, and we withdraw sexually or emotionally to shield ourselves from rejection.

By hook or by crook, though, partners do adapt to one another's body image, although not always in the healthiest ways.

EIGHT

Couples

When two people join together as a couple, they bring with them their relationships with their parents, siblings, friends, and ex-lovers; their psychological needs; and the images they hold of their bodies. That last item is of no minor significance in an age in which prospective partners meet at health clubs and diet centers and through personals ads.

If partners come from different backgrounds, a shared set of beliefs about the body can sometimes substitute for common interests and experiences in other areas of life. Remember Carrie and her husband, Bob? They're both teachers but have little else in common. Carrie is fourteen years younger than Bob; her family is Jewish, while Bob's is Protestant. Carrie grew up in a city, Bob in the country. She likes classical music, he likes rock 'n' roll. They don't even share the same temperament. He's steady and calm, she's moody and excitable.

But they both exercise like crazy, and the lone bookcase in their living room is filled with the last ten years of *Prevention* magazine; and so Carrie could say (and Bob echoed the sentiment): "We're very similar. We both do the holistic approach to health. We both work out, we get our heart rates up, we eat extremely well. We don't eat any meat. I won't say I'm a vegetarian, because we do have fish once in a while, but I don't eat any red meat or poultry. We're very aware of whole grains."

Over dinner, Carrie and Bob talk about stress fractures and whether to add another B-12 pill to their daily stockpile of supplements. They respect one another's devotion to exercise and refusal to eat meat, much as devout fundamentalist partners revere each other's study of the Good Book and abstinence from alcohol.

In other relationships, just the reverse occurs. Conflicting beliefs about the body act as a wedge, despite a shared background and common interests in other areas. I witnessed this in action at the suburban LA home of Madeleine and Sean, a husband and wife who, like most couples in America, are homogamous in the traditional sense—they're from the same religion and social class, they grew up in the same city, and they're just a year apart in age.

Madeleine, a prominent talent agent, and Sean, a cardiologist, have great respect for each other's professions, and they work well together as parents. But the physical differences between them are marked: where Sean is very thin, Madeleine is heavyset, and while Sean exercises every day, Madeleine almost never does.

In fact, Madeleine's day begins at five in the morning, when Sean leaves their bed to jog five miles. She tries to persuade herself she is still asleep until six, at which time she gets up to feed and dress the two children, eight and eleven, before the bus from their private school arrives at seven. In the kitchen, over her first cup of coffee, she generally experiences her first guilt pangs of the day. *I should run too,* she thinks, feeling fat and listless as she puts the kids' cereal on the breakfast table, *or at least ride the stationary bike.*

When I interviewed her one Wednesday morning, she told me straightaway, "I'm going to turn forty-five on Sunday, and I've never made time for athletics. When I went to school girls did nothing. In gym we were supposed to stand around and look cute for the boys.

"It's different for my daughter. She plays soccer. She's very fast on her feet, and she looks terrific. She has lovely long thin limbs. It's still amazing to me, an intellectual. When I was in grade school girls went into music and art, not soccer.

"I had a terrific body from the time I was twelve or thirteen, clear into my twenties." She shrugged. "Very feminine, very curvaceous. I matured early—I was very sexy—and I was proud of it. I had a great bustline and I was famous for my tiny waist.

"Now I look *awful* in a bathing suit," she went on energetically, as if this self-criticism felt therapeutic, still without a single question from me. "I have that extra little tire around my waistline. I'm not able to move as quickly, it's annoying to have gone from size eight to twelve. I can't stand it, looking at that big round stomach in the mirror."

Listening to her in the living room of the family's ten-year-old ranch home in Sherman Oaks, as she went on to find fault with several other parts of her anatomy, I received the impression, despite her disclaimers, that she rather *likes* being heavy—that she enjoys having some weight to "throw around." So when she finally ended her soliloquy—"Let's face it, I look generally dumpy"—I asked if she would like to change her appearance. She answered not with her own preference but with society's. "I have a good fifteen pounds too much on me. I weigh 140, and I ought to weigh 125, according to the insurance tables and all that. I'll never weigh 115, as I did in high school, but at least I could get down to 125.

"Sean maintains we all ought to weigh what we did in high school, that there's no reason to gain weight every year. I'm sure he's right, but then I've had two children. I never got muscle tone back after my children. I gained a lot of weight in both pregnancies and had very difficult deliveries, and I breast-fed for nine months both times. At that time the doctors said you were supposed to gain twenty pounds. Now they say any gain is horrible. Who knows which is correct."

She sounded angry when she said this, and I told her so.

"I look at my mother's generation," she responded immediately, in a more measured voice. "My mother was a lovely-looking lady at forty, but she had flab on her upper arms. She didn't exercise. She had a very vital life, working full-time, and she was a terrific mother and a terrific hostess, but she sure didn't worry about what she looked like on the beach.

"She's eighty-three, by the way. The lack of exercise did not kill her. My father was tall and thin, very well-built and handsome, and I never once heard him say a word to my mother that she should change her appearance. All he did was say how lovely she looked.

"I resent it that she was allowed to look forty-five at forty-five, but I'm supposed to look twenty-five." The ire returned to her voice. "There's something crazy about expecting me to make pancakes for everybody on Sunday morning and not eat them. And that's the only way to keep trim, to have dry toast and half a grapefruit, and exercise. My son's birthday was last night. It was my role to make a beautiful home-baked chocolate cake. My daughter's birthday is next Thursday. She wants a lemon cake.

"My work also involves a lot of dinner parties and restaurant meals. Tonight we go out with friends, Sunday we go to a big party one of my clients is throwing—I have to bring a dessert. Next Tuesday night I'm going to an extravagant dinner party. All of this is in direct contrast with what I need, which is Lean Cuisine and a pear for dessert."

I rephrased my earlier question: Would she prefer to conduct her life differently for the sake of her body? Again she invoked the experts.

"I read Jane Brody," she said, respect creeping into her voice. "Everything she says about watching your diet and eating complex carbohydrates and getting rid of fats is very sensible, but I'm not one of these people who is going to raise my suburban children to eat wheat germ and soybeans, and I can't be rude at parties.

"So the only choice is to exercise the pounds off. I hate being indoors all day long, I need exercise, I accept that. But where do I find time? Of course, my husband says, 'You can make time for it, look at me.' He's a busy doctor, and he made time for it."

In an uncharacteristically quiet voice she added, "He's in shape, women comment on that, there's not an ounce of fat on him.

"But truthfully," she said cautiously, voicing her own opinion for the first time, "I don't think his upper body and his lower body match."

I laughed spontaneously, having known joggers in Sean's age group with legs like tree trunks and arms like pencils. The younger generation of male runners may also pump iron, but joggers in their forties and fifties often do nothing other than run (and run and run).

My amusement freed Madeleine to say a bit more.

"I like a man with a little extra meat on him in bed," she went on, smiling a wicked smile. "I like sort of the Arab-sultan look. It's comforting to have some bulk. You feel protected. I don't want a fat man, but a little extra weight on a man is attractive."

Madeleine knows she is *supposed* to like Sean's body. It's healthy to run and keep yourself fat-free. Other women, with heavier husbands, openly envy her. But, as she went on to tell me, the costs to her are great, and she and Sean are out of sync.

"When he first took up running, he'd run anytime he could," Madeleine recounted, "when he got home, on the weekends. He's a doctor, so he works on Saturday and Sunday mornings. Then he'd come home and run for the rest of the afternoon.

"Men see it as their God-given right to go and exercise. For the woman to do that means hiring a babysitter and making all those calls and arrangements to actually go out of the house. I put my foot down, so now it's mostly confined to five A.M., except when he runs marathons.

"He gets up at five to run, and then he's working all day and gets home late. So he has no energy for sex. I always have more energy than he does, and I rather enjoy sex, so I miss it."

The first ten years of their marriage Sean and Madeleine were very close, but in the seven years since he's been running, she's noticed a growing remoteness. It's hard to say whether the emotional distance is an effect or a cause of the running. But what certainly *is* true is that Sean's running is the occasion for battles over power in Madeleine and Sean's relationship.

Several days after I met with Madeleine, I interviewed Sean. While Madeleine talked to business associates and friends on the telephone in the kitchen, Sean and I talked upstairs in his little study, furnished with hand-me-down furniture from their living room, a desktop IBM-PC, and stacks of medical journals and issues of *Runner's World*.

Trim and long, tanned and very slightly balding, Sean looked the prototype of the suburban California émigré male. Though it was a cloudy, cool January evening, the top two buttons of his Pierre Cardin sports shirt were unbuttoned to reveal his hairy, undeveloped chest.

He didn't hide his feelings of physical superiority.

"It's hard not to become somewhat self-righteous about it," he said. "I think what I'm doing is correct from both a medical and a personal point of view. The medical evidence is pretty convincing that vigorous exercise is clearly something that everyone should be doing. It's not a proven fact, but from a medical point of view I think that an hour running is the best expenditure of time you can make."

When I asked if he encouraged Madeleine to exercise, Sean responded, paternalistically, "As a physician, I see it as a medical necessity, but I don't nag her with, 'You're going to die, you're going to get a heart attack or stroke.' I think my concern comes from knowing Madeleine would feel better, have a better self-image were she twenty pounds lighter and exercising. Exercise reduces stress, and that would be a marked benefit to Madeleine."

Actually, it's not so clear that exercise would serve to reduce stress for someone as hard-driving as Madeleine. Beyond that, I wonder if her health was actually the primary consideration behind Sean's wish for a fitter wife. At several points in our discussion he made remarks about how much more attractive she *used* to look and about the temptations to infidelity he faces at work and at his runners' club.

Though he may yearn for a younger-looking wife, another part of Sean clearly doesn't want Madeleine to get in shape, because his fitness is his claim to fame in their relationship, both in the minds of those who compliment him (and her about him) at parties and in his own sense of himself.

His father was a wealthy Jewish physician on Manhattan's Upper East Side, and as a child Sean learned more about opera than about sports. "I never thought much about my looks," he said. "I never thought I should weigh less or be athletic. I guess I knew I was reasonably attractive. I was very conscious of looking and dressing nicely, of making a good appearance, but I never wished I was taller or anything like that."

Sean started running seven years ago, when friends began remarking on his belly and he found himself within arm's reach of forty and huffing and puffing while mowing the lawn. First he took up tennis again, which he'd played twenty years earlier in college, but it didn't fit into his schedule as a cardiologist frequently on call. "When I was in the Air Force there were several captains about fifty who were obviously in far better shape than I was," he explained, "and I've always remembered them. In the back of my mind I often thought, 'Gee, I'm never going to end up at age fifty looking like any of these guys unless something happens.' I was never athletic. Most of my life I was on the heavier side. Never obese, but I liked to eat.

"A close friend said he had started jogging and I really should too. I'd looked at people jogging and thought it was stupid, but I guess when this friend suggested running, somehow it rekindled that image in my mind of those guys who were on the verge of retirement and in terrific shape. They were thin and healthy-looking, the kind of guy you would really like to be."

No doubt this account of Sean's was truthful, but something else happened at the time Sean became a runaholic: Madeleine's career took off. After she had worked several years in a large talent agency, representing young actors and actresses who would never get beyond the soap-opera circuit, one of America's foremost stars asked her to represent him. With the proceeds from this client's next couple of film contracts, Madeleine opened her own agency from home.

Sean's career, meanwhile, had stalled. He was still at the unglamorous suburban hospital where he'd gone after finishing his residency (he's there to this day). He wasn't learning anything new, and none of the prestigious practices was begging him to join.

And so Sean considers exercise his personal achievement, something he's proud of, something that's his alone. For all the times Sean has communicated to Madeleine his wish that she'd exercise and lose weight, he's done nothing to facilitate it. "I know couples," said Madeleine in my second interview with her, "who run together, who play tennis together, they play doubles on weekends and all that. But I'm stuck with a man who will not do it with me because he's so much better and doesn't want to waste his time. He won't play tennis with me because he says I'm not good enough, which is perfectly true. He would never run with me because I can't keep up with him. The only thing we ever did was go walking in the mountains together last summer when we took the kids on vacation. All I ever saw was his rear end. I was half a mile behind him."

In the week since we'd talked, she had "been rankling over this exercise matter," she told me, and she was ready to lash out about it. "Does Sean do anything," she went on, "to make it possible for me to have time for something athletic? No. *I* go with my children to arrange their music lessons and their soccer, *I'm* the one who picks them up from school. *I* cut my workday short to do all that."

This, too, is true, and yet it doesn't quite explain Made-

leine's and Sean's physical disparity. Their housekeeper
drives and could easily transport the kids. Rather, the point
is that Madeleine, like Sean, has a strong sense of her own
special identity. If his self-esteem derives from being fit,
hers depends, ironically enough, upon being a bit out of
shape. Madeleine views herself as someone who's too busy
for exercise.

"Women these days are expected to be successful mothers,
successful wives, successful homemakers, successful in our
careers. By all standards I'm successful at all of those, but
it takes a tremendous amount of time. It means I spend
time with my children every day, which means I read scripts
every night past midnight. It means lots of office work and
meetings in Hollywood. And then to *also* fit into that looking
like Jane Fonda . . .

"It's a matter of priorities. Exercise is not a natural priority
for me, and I resent it that I'm actually thinking of cutting
my piano playing, which I do every night for a half-hour, to
sit on the exercise bike Sean bought me, which gives me
no pleasure whatsoever. Playing the piano gives me tremendous pleasure. I took piano lessons for ten years when I
was a child. Now I take lessons from a marvelous teacher
every Saturday afternoon. It is my personal indulgence, and
I feel a real sense of accomplishment. I play totally for myself."

Exercise has just the opposite effect for her. The times
she's tried athletics, she's experienced only failure; and later
in our discussion, while Madeleine is talking about how her
friends all live thirty or more miles away, it comes out how
her flabbiness serves as a mark of status in her own mind.

"Most of the women who live in this neighborhood have
a different life-style from mine," she scoffed. "They play
tennis and golf all day, and they take naps in the afternoons.
I can see the utility in that, because they have gorgeous

bodies at my age. That's their full-time job, looking good, and shopping and instructing the maids. And taking their anti-depression medication, which they get from the psychiatrists they see three times a week."

I wonder what would happen if Madeleine lost weight. If she became lean and fit, would she and Sean be happier together? I doubt it. More likely, they'd be at each other's throats. As Richard Stuart and Barbara Jacobson point out in their book *Weight, Sex and Marriage*, most husbands of overweight women say they'd like their wives to lose weight, but many don't really mean it. Stuart and Jacobson, who are marriage counselors, describe husbands who urge their wives to slim down but then sabotage their efforts by grumbling about the low-calorie dinners they're forced to share or their loneliness when their wives go off to exercise or diet classes.

There are several good reasons, say Stuart and Jacobson, why a husband might secretly prefer that his wife stay plump. For one, he need not feel as guilty about his own faults and limitations. He can blame his poor sexual appetite or performance on his wife's lack of sex appeal, for example, without worrying that she'll have an extramarital affair. As long as she's fat, he reasons, other men won't be interested in her. In fact, Sean and Madeleine both complained to me that they hardly ever have sex. She blamed it on his 5:00 A.M. runs, he on her unsexy extra pounds.

Accommodation

Although Sean urges Madeleine to exercise and lose weight, I suspect it would upset the balance of payments in their relationship were she actually to do so. As things stand, she has the glamorous and successful career in the family, and he has the glamorous and successful body.

With fitness and thinness such prized possessions in our society, husbands and wives bargain with one another about rights to them just as they do about money or time. Some partners grant one another carte blanche to work out and eat lean (Carrie and Bob are examples). But in many marriages, the agreements about what each partner is allowed or required to do for his or her body are complex, tied in as they are to a system of tradeoffs in other areas.

In a suburb northwest of Chicago I met a man named Don, thirty-two, who looks like Beaver Cleaver grown up— plain, round-faced, and tubby—and his wife Louise, thirty-one, also rather plain, but well proportioned and not overweight. Don told me, in not so many words, that he wishes his wife would get out of shape. What he said was, "I think she's gone overboard with this exercise and healthy-eating business." He also said he finds Louise more cuddly when she weighs an extra fifteen pounds.

I'd interviewed Louise before I talked with Don, and she'd made it clear that she has no intention of fattening herself up or of cutting back on her exercise program. Since marrying Don, she's accommodated herself to him in several other ways, but on the matter of physique, it was Don's, not hers, that she felt needed to change.

Louise and I talked in the kitchen while she prepared dinner for their two-year-old son, and the topic of her weight came up right away. "I was a little heavy," she replied when I asked how she felt about her body when she was growing up. "Not overly heavy, but not thin. On the high side of normal, I guess you'd say." What about now, I asked. Is she happy with her figure? "Pretty much," said Louise, who was wearing gray sweatpants and a sweatshirt. "I'd like to lose ten pounds, but wouldn't we all?

"I had a big weight gain when I was pregnant," she went on to admit. "But I was able to get down to a lower weight than before within eight months, which I felt real good about. Now my husband thinks I'm too thin, which I think is crazy. I tell him we *both* need to keep to a diet. He needs to get rid of twenty pounds."

I asked if she would find him more attractive if he were thinner. "It's not so much for my physical attraction to him that I want him to lose the weight," she answered. "I just think he'd be a whole lot healthier ten or twenty years from now if he kept his weight down, because he does tend toward high blood pressure."

She turned away from me at that point and began cutting carrots. "Since I quit my job and have been home full-time," she said, "I've tried to make it a project to prepare more balanced meals. I have more time to read about that kind of thing now. I know Don would rather come home to a cheeseburger—that's his favorite dish—but I'm weaning him off all those calories and cholesterol."

Her staying home was Don's idea, she indicated, and she's still not sure she did the right thing in giving up her career with an insurance company.

"After you have your first child, it's not easy deciding whether you should continue your career or stay home and go to exercise classes and get yourself back in shape and have another child in a year or two.

"Don felt very strongly that I should stay home. And I think it's important to the baby's development that I be there. Some days I'm drained by reading him his *Go Go* book six times, but I don't think a babysitter, who's there just for the money, would do that. And I think it's important that he be read to as much as he's willing," she said with conviction over the noise of the blender in which she was pureeing the carrots.

"If you spent a lot of time in the work force and built a

career up for yourself, and you don't dislike your job, it's a major change to just say, 'I'm not going to work anymore.' Mentally it's a change because you're used to stimulating your mind. And physically it's a big adjustment also. You see a lot of the new moms out here who weigh more than they should. It's real easy to let that happen when you're home so much. The only people you see some days are your husband and your child."

Louise turned off the blender, then added: "And I've had my own income since I was ten years old and babysat, and all of sudden I don't have a paycheck, and there's a psychological, you know, *rift* there. It makes me feel a little more dependent on my husband."

As she poured the baby's food into a saucepan, I asked if her daily life was typical of women her age in this suburb. She said it was, and described her routines. "My friends are all about my age with toddlers at home, and we have a real little daytime life here. I go to the park and make friends. I have a play group and I belong to Junior League and I do community work and volunteer work, which is important because otherwise you sit around all day at home in your jeans.

"I have friends I can talk to here," she repeated, as if to be sure she'd made that point clear. "A lot of times we'll talk about problems with our children or the fact that, gee, we used to have these glamour-type jobs and we don't anymore, but our husbands still go to work; a lot of women want to vent that. It's nice to be able to admit things like, 'Financially we went from two salaries to one, and I wish I could go to Europe this year.' And it's nice to have someone say, 'Here's what we can do. We'll have a Europe party one afternoon, and we'll look at each other's slides and drink some French wine.'"

I asked what she enjoys most about her life, and Louise

gave an answer that reveals another meaning of exercise for the modern woman: "I enjoy the exercise program I take three days a week and on Saturdays at the health club. It's not real strenuous—I just enjoy doing something that's for *me*. I feel I should get a gold star on my chart just for going," she said with a little giggle. "Instead of doing everything for my husband or my child, I have that forty-five minutes of exercising just for me. And then I can take a shower at the club and not have to be worried about my shower being interrupted because the baby has to have his diapers changed or something.

"Excuse me," she said then, noticing the baby's food getting cold on the counter. At that moment Don walked into the kitchen holding the baby. He shook my hand and asked Louise if she was ready to feed the child.

It seemed strange to me that Don would appear just when he was about to be summoned. After I interviewed him, though, and learned how insecure he is, I concluded he'd probably eavesdropped during my talk with Louise. He turned on a radio in the kitchen to ensure that Louise couldn't hear us and took me down to the basement for our interview.

Don still feels nervous and guilty about having insisted that Louise quit her job. It's unclear these days whether a husband has the right to make those demands, but more to the point, her staying at home entails major obligations for each of them. She's expected to find fulfillment as a homemaker, and he's expected to bring home the bacon and then eat the low-fat food Louise cooks.

Harder still, Don has to be The Man of the Family, now that his wife has taken on a traditional female role. As a boy he was overweight and slightly effeminate and always did poorly in gym class. He became a somebody through the local radio station; powerful-sounding and in a job people think of as sexy. "It's tremendously satisfying," he told me at one point, "to be twelve years out of high school and go back to your hometown and see that the football quarterback and his macho buddies are literally pumping gas. Because you knew there was a reason you were applying yourself back then and putting up with their making fun of you."

Don has lived in fear and envy of the models of proper masculinity he grew up with and still sees around him. At five-eleven and 200 lumpy pounds, he hasn't been able to forge anything resembling the ideal in his body, but he *has* achieved it in his voice. Through diligent work Don has become, on the air, the ideal male. He talked to me about the technicalities of that process every time he got a chance in our discussion.

"I have a heavy voice, which was a problem at first," he said, lowering it an octave to illustrate. "I sounded more pompous and bombastic than I should have sounded—like 1940s radio, hard and heavy, like the newsreel announcers. In the 1980s we want to sound conversational and bright. I consciously worked on lightening my sound, using more of the higher part of my voice and more inflection.

"Then for a while I sounded *too* light. I listened to tapes and I was sounding nasal, like I was talking through my nose, very wimpy. And I certainly don't want to sound like a wimp on the air, because you need to sound authoritative and believable. Now I think I've mastered the right mix."

Still, a man's Adam's apple is a rather shaky place on which to rest his masculinity. Don is awkward off the air and doesn't have many friends, even though he's a big-city radio announcer on one of America's leading AM stations. When I asked if he exercises, he confessed that he never uses the health club membership he bought because he's embarrassed to undress in front of other men.

In the work world, he's relied on Louise to push him forward. At the time Don and Louise met six years ago, what might be called her masculine side was attractive to him because he needed to borrow some of it. He was an ambitious guy who couldn't seem to get himself a better job than at a mediocre station in the suburbs. "People perceived me as aloof," Don explained, "because I've always hated small talk, and so rather than make small talk, I'd make no talk. I like working in radio because I appreciate an economy of words, and in radio you say things as quickly as you can. Whereas Louise was in sales and didn't believe in economy of words. She's always been a glad-hander, and I think that's been good for me. She was constantly telling me, 'You get up there and shake their hand, because they'll never notice you otherwise.' "

But once he mastered both his reticence and his voice and moved to the big time, Don needed another version of his wife, one that took some pestering to produce. "I need all my energy just to feel comfortable in my job, I can't be worrying about whether the baby is being cared for properly or whether dinner is getting made," Don let on. "I came here two years ago worshipping the station, and I was intimidated by the people I worked around, which was bad because it got in the way of my being as good as I could be. That's something I'm just now overcoming."

In Don's line of work, fear is not only a personal problem but a threat to his livelihood. If he doesn't feel properly detached and confident, it's hard to sound authoritative on the air. He holds himself back and stumbles over words.

"This is a business where you're only as good as your last show, and if you come home after a bad show you're going to be mad for a couple of hours. Louise has learned to accept that."

She's had to. The linchpin in Don and Louise's relationship is Don's masculinity, because so much they both care about depends on it. He makes twice what Louise did when she was working, and so when he's feeling weak and insecure and consequently fumbles on the air, the family is in danger of losing its life-style. And conversely, if he performs well and is seen as a ratings improver, other stations will bid for him. In that case, Don and Louise could relocate, in time for their son's elementary school education, to the larger house and fabulous school district of their dreams.

Like many men who feel vulnerable, Don has struggled with his wife to change her from a self-sufficient career woman into a more traditional woman: a caretaker and mother. Except for the fact that she doesn't look matronly enough, he's succeeded. At the same time, Louise has been reshaping Don to look and act like her notion of an attractive, healthy husband.

Psychologist Robert Hess and sociologist Gerald Handel have written that, in any family, each member holds an image of every other member, and also of himself or herself—images that are "compounded of realistic and idealized components in various proportions." They suggest that if a family is to flourish over the long haul, "the images that members have of the family and of one another must in some sense tend toward compatibility."

So it's not surprising to find husbands and wives trying to realign one another's body-images. But where many partners err is in assuming that in order to be compatible they must be identical. "The issue involved here is not one of how similar the members must be to each other," write Hess and Handel. Rather, "the issue is whether the differences and similarities among the members are mutually acceptable."

In other words, spouses need not be equally fat, or equally fit, in order to have a happy union, contrary to what Don, Louise, Sean, and Madeleine believe—and contrary to what some self-help books recommend.

Chris Pepper Shipman's *I'll Meet You at the Finish!*, for instance, preaches that couples who exercise together stay together. Without bothering to place her tongue firmly in her cheek, Shipman describes how her southern California marriage turned marvelous once her marathoner husband converted her to running.

Multibillion-dollar industries live off people who hope to help their marriages, or to recover from divorce once their marriages fail, by altering their bodies to match the images their would-be lovers hold for them. Such people are a primary market for health club promoters, weight-loss programs, cosmetic surgeons, and others who earn their living by remaking other people's bodies.

PART III

REMAKING THE BODY

NINE

In the Name of Health

For better or worse, we Americans have become obsessed with our bodies.

Some say such an obsession is clearly for the better. Health promoters point out that we're exercising more, keeping our weight down, and eating more wholesome meals. But critics say the current emphasis on the body is making us into a nation of shallow, addicted narcissists who care about nothing beyond our own gratifications.

I say both views are based on the same faulty assumption: that an interest in the body is entirely a self-centered act.

It is for profoundly *social* reasons that we've directed so much attention to our bodies in the 1970s and 1980s. The body boom may *look* individualistic, but in this case looks are deceiving. To be sure, the decision to have cosmetic surgery or to lift weights is made by an individual, and some of the pathologies of the age, such as exercise addiction and anorexia, are borne by individuals. But changes in society are very much implicated. As women's roles shifted, eating disorders became more common; as America sought to regain its might after the Vietnam War, the musclebound look grew popular.

What's more, when individuals set out to improve their bodies, usually they opt for social, not solitary, means. They enroll in health clubs or weight-loss programs. Or they develop relationships with professionals—exercise instructors, diet doctors, plastic surgeons.

And people decide to alter their appearance not at indiscriminate points in their lives but most commonly when their social circumstances are changing.

"Women come to me at pivotal moments," said Allison, a beauty-makeover specialist I interviewed in New York. "Probably half of my clients are going through divorces. They're ready to give off a whole new set of signals, but they need a little help making it work. They're getting into circulation again and have visions of becoming glamorous and sexy. Or they're going more into their careers and need a bolder look."

Other clients of Allison's seek revision rather than metamorphosis. "A lot of women," she reported, "will say, 'I got married and I had a kid and I dropped out for a while, but now my son's a toddler and I'm getting a part-time job and I want to update my image.' Or sometimes college students will come to me when they're about to graduate, to professionalize their style before they go into the job market."

She's also worked with women who use her services as a *stimulus* to change: "They feel stuck. They look in the mirror and they feel ugly. They hate their bodies and they hate their clothes. They may love their babies and their husbands, but they're fighting with them too, because they feel taken for granted. They want to do something to effect a change, so they come to me to make them feel better.

"Once they look better they act like different people. They get along better with their kids, and their husbands start noticing them again."

Some people alter their appearance to test whether they

are capable of new challenges. "A lot of times the makeover is a springboard that leads to other accomplishments," said Allison. "They figure if they had the nerve to go through with the makeover, they can go out and get a new job too . . . or dump the man who's been mistreating them. They get positive feedback from others, and it gives them the nerve to do other things."

Cosmetic surgeons give similar reports. According to a study from Harvard Medical School, a substantial percentage of patients choose to have cosmetic surgery for reasons they are unable to face at the time but which reveal themselves later. Soon after the operation, they make a major life change, such as filing for divorce.

More Than a Haircut

Some body remakers are acutely aware that they provide social services to their clients. An electrologist named Karen, whom I interviewed at her office in Greenwich Village, sees many of her clients regularly for months on end, and frequently finds herself playing social worker on their behalf.

"For example," Karen explained, "one of my clients will tell me she and her husband are moving to the suburbs. Then another will come along who's just gotten divorced and needs a smaller place to live. So—to protect the confidentiality of my clients—I'll tell the second woman that I was at the health club yesterday and somebody was talking about giving up her apartment at such-and-such a location.

"Or somebody will be bitching that her younger brother the accountant never gets any dates. I'll give her the phone number of one of my clients—and I have *loads* in this category—who is a single woman looking for a stable man. To be discreet, I'll say she's in my aerobics dance class."

Karen described her clients as women on the make in the

business or theater worlds, who go to her (and to plastic surgeons, personal trainers, and weight reducers) because they don't want anything unsightly to stand between them and a six-figure-a-year income. But they get much more for their $35 per half-hour than just an electrocution of their hair follicles. In addition to her networking service, Karen listens appreciatively as they relate their most personal psychological, marital, and financial problems. She's heard everything from minute-by-minute accounts of secret sexual rendezvous to the intimate details of insider stock trades.

It's common knowledge that electrologists, hairdressers, manicurists, and masseuses function in our culture as surrogate psychiatrists and trusted friends. The usual explanation is that a majority of the population either can't afford or doesn't know how to locate appropriate confidants. In many marriages and friendships, unspoken rules prohibit "heavy" talk; and with work colleagues the dangers of revealing oneself may be great.

The body remakers I've interviewed, including Karen, give a second reason why people tell all to their hairdressers. They say that a person feels a kind of intimacy toward someone who has been allowed access to his or her body. After all, the body is the most private and personal of possessions; once you've turned *that* over to a stranger, the unloading of a few skeletons from your closet is no big deal.

I think there's a still less conscious reason as well. On some level, we envision these people as having special powers. Rationally, we know they can do no more than sharpen our appearance, but *ir*rationally, we believe they can change our lives. We imagine them capable of making us into—in the words of an advertisement for a cosmetic surgery center in Minneapolis—"some body new." Obviously (so reasons our subconscious), anyone who can transform us from one type of person into another must be able to readjust us emotionally as part of the package.

Karen, unlike the vast majority of body remakers, actually has some training in counseling. She was a psychiatric social worker, serving as a nurse at an inner-city psychiatric hospital. She'd spent half a dozen years in the place, was on the verge of turning thirty, and felt herself slowly simmering into a full-fledged burnout. Most of the patients, poor and suffering from schizophrenia, had been through the system many times. The best a staffer could hope for was to make it through a whole week without violence.

One day in 1981 a well-dressed but terribly agitated woman was admitted to the facility. "Give me a razor! Give me a razor!" she shouted over and over. Karen ignored her at first, but when she didn't let up after an hour, she went in to talk to her. "She was sitting on the edge of the bed," Karen remembered, "and I noticed she had a whole chin full of hair. She'd obviously been in a psychotic state for a week or so and had stopped taking care of herself. By the time she got to the hospital she was starting to come out of her psychosis, and she must have caught a glimpse of her face in a mirror.

"I was ahead on my paperwork for the day, so I just went to the supply cabinet, got a safety razor, took her to the bathroom, and shaved her chin.

"The minute the hair was off, she was a different person. She relaxed, she spoke coherently. My supervisor gave me hell about it, said that I'd taken a terrible risk, but it was probably the most effective therapy that woman had had in years."

Hair removal will not often bring a highly disturbed patient back to earth. But more than a few progressive psychiatric hospitals employ full-time cosmetologists. After all, an important step for any of us when we move from a private sphere to a public one is to dress and groom ourselves accordingly.

As for Karen, the incident helped her decide to become an electrologist. She'd been considering the idea for some months, ever since it had been put to her by the electrologist she herself was seeing twice a week.

Karen had met this electrologist after what she refers to as the most embarrassing night of her life. "I'd curled up in bed with a man I'd just made love to," explained Karen, "and he put the sheet in between us. I had hair on my stomach and I used to shave it. I guess I must have been scratchy that night. Well, I was mortified to see his reaction, and as soon as he left my apartment the following morning, I picked an electrologist out of the yellow pages and made an appointment."

Karen became friends with the electrologist, who helped her get training and eventually took her on as a partner. After a few years Karen opened her own practice, which has thrived from the start.

Weighty Business

The manner in which Karen was recruited into her new occupation is not unusual. Professional body remakers quite commonly come from the ranks of those who've suffered personally with the problem they now treat. In some cases, having been a sufferer is actually an official prerequisite for employment.

This is particularly common in the weight-loss industry, as I learned from a woman who owns a Weight Watchers franchise in the Midwest that runs 400 meetings a week and employs almost 500 people. Nearly all of those employees are "lifetime members" of Weight Watchers, which means they used to be overweight.

"We're in a business in which weight is essential to the integrity of the organization. We wouldn't have a ghost of a chance in the community if we didn't present proper role models ourselves," said Eve, who sets aside time at every business meeting in case any members of her staff are experiencing weight problems they want to discuss.

If an employee does gain weight above the prescribed range, Eve prohibits the person from official contact with the public, sets up individual counseling with an experienced group leader or someone from her managerial staff, and requests that the person attend a Weight Watchers group until the necessary pounds are lost.

As of the time I met her, she had never had to fire anyone because of weight. Those who'd been unable to shed the required number of pounds had resigned on their own. But she worried she may someday be sued on grounds of discrimination. "It's a very sensitive issue around the office," she said.

Eve, who wears her long black hair in a Gloria Steinem cut, is forty-three, close to six feet tall, lean, and herself a lifetime member of Weight Watchers. "My weight problem started when I was twenty," she told me early in our dinner at an elegant restaurant. "For the first ten years I handled it by starving part of the time and bingeing the rest. By the time I turned thirty, I couldn't do that anymore. I couldn't go without eating because I'd get dizzy.

"So I started putting on a lot of weight. I felt like an elephant. I tried everything. I went to doctors, I tried the liquid protein diet, which made me lose my hair. I took diet pills. Each of them worked for a little while, but not very long. I was on what we call in the weight-loss business the yo-yo syndrome, which can be more detrimental to the human body than carrying extra weight.

"Eventually I joined Weight Watchers, and believe me, I was one of the biggest skeptics who ever walked into the room. Number one, they were telling me to eat food, which I figured couldn't possibly work. If *not* eating wasn't working, how could eating work?"

Rather than counting calories, Weight Watchers provides menus which include precise portions of fruits, vegetables, dairy products, and meat, fish, or poultry. The member is required to eat three substantial meals a day and a snack.

"I was skeptical, but I promised myself that I would do what they told me," she went on, narrating a story she'd no doubt recounted many times in the ten years since she lost twenty-nine pounds, but one she told with obvious relish. "So I got the program, went home, and for the first time in my life, I cooked. I'd never been in the kitchen before, but here I was taking pains to cook properly. I weighed everything as they told me to, and I ate everything they allowed me to eat, which was volumes more than I'd been eating."

At this point we were interrupted by the waiter. We both ordered the broiled swordfish and mixed green salad, and Eve continued her story.

"They don't allow you to weigh yourself between meetings, because if you lose weight you'll reward yourself with food and gain back everything you lost. So I showed up the next week absolutely *certain* I'd gained five pounds. I was so embarrassed. I went to the meeting in my raincoat. Actually I'd *lost* two and a half pounds.

"I thought it was a fluke. But the next week I lost again, and the next week again . . . and I got hooked. It gives you something to look forward to, going to the meeting and weighing in.

"People who haven't had a weight problem just don't understand," she continued. "They look at the person and they think, 'How could she let herself get that way?' They have pat theories about why people are fat, like assuming that people overeat when they're depressed. Well, sure, some people eat when they're depressed. And some people eat when they're happy. Some people eat when they fail, some people eat when they succeed. There are all different triggers for overeating."

What's *her* answer to the question of why people get fat, I asked.

"Interestingly," she replied, "I've never been asked that before." She took a sip of water while framing her answer. "I imagine there are as many reasons for someone to be overweight as there are"—looking around the room for a good comparison—"recipes for bread. *My* reasons for having the problem were all emotional. Every emotional high and low caused me to eat.

"Also, as a child I was constantly given rich food. Ice cream, candy, cakes. My parents, whom I love very much, weren't doing this maliciously, since I was reasonably thin at that time. But as I got older and started gaining weight, I didn't know how to eat."

Members of Weight Watchers don't make this mistake, she told me. They learn proper ways to eat and pass these along to their children.

Eve went on to describe what she called the "magic" of the program: the newspaper she publishes every month with "before" and "after" pictures of members who lost thirty or forty-five or eighty-five pounds and look like different people; the heartwarming letters she receives from people whose lives have been turned around.

Actually, Eve didn't need to sell me on the value of the self-help method; I consider it one of the great social inventions of this century. These groups are so successful that professionals ask *them* for assistance at times—as evidenced by Alcoholics Anonymous chapters within psychiatric hospitals.

The key to the success of self-help groups is an ingenious bit of sleight of hand they perform. They treat members' weaknesses as assets. Instead of feeling guilty or hopeless because you went off the wagon or into an ice cream parlor, you're encouraged to relate those experiences at group meetings so they might help others.

And if you do succeed at controlling your weight, you may even gain a job. Eve informed me that over the years her franchise has hired as group leaders and clerical workers about 1,000 members of the organization, many of whom had been depressed and unemployable housewives before joining Weight Watchers.

By bringing together people from diverse backgrounds, groups like Weight Watchers accomplish another remarkable feat of social engineering. This Eve called to my attention when I asked her—after the swordfish arrived—to describe the ingredients in an effective Weight Watchers meeting. She immediately pointed to the fact that groups in her territory include blacks and whites, Jews and gentiles, rich and poor, young and old, women and men.

"Newcomers walk into that room with a lot of pain and guilt, and what do they see? They see they are not alone. Even though their problem may take on a different tone, all the people in that room, regardless of how much money they make or how wonderful they look now, have the same problem.

"It becomes a real family in that meeting room. You don't sit in a meeting for an hour and a half and just think about how many pounds you've lost. You start to listen. You go in thinking you're unique, you're the only one who has starved and not lost weight, who can't control her appetite. Then you hear the other people, and you feel all the support there is among people who have a common problem."

Another reason why a "family" feeling develops in these sorts of groups is that people who lose weight often discover that their real family and friends are less than supportive. When someone commits to change, friends are likely to feel threatened. Eve, who led groups herself for several years, told me that a great deal of time is spent in Weight Watchers meetings discussing the reactions of family and friends.

"People who lose weight," she said, "do more than lose weight. They go through a transformation. Very often, those you consider your closest friends don't like seeing you looking that good. They tend to get jealous or angry." Eve estimated that about half the members who lose weight and keep it off also lose a close friendship in the process, or a marriage.

Yet the problems that members experience in their personal lives are off-limits in Weight Watchers meetings. Eve said her group leaders are instructed to stick as closely as possible to the topic of how to eat properly.

"We never get away from why we're there," she said, wagging her fork at me as if reprimanding one of her staff. "We're not in the psychology business. We are not therapists. A Weight Watchers meeting is not a therapy session. It is not a bull session, and it's not a gripe session."

What if someone comes in and says, "I had a big fight with my spouse, so I had a bad week," what do you say to that person? I asked.

"What did you do?" she said to me as if I were the member in question.

I overate, I said.

"What did you eat?"

Two chocolate cream pies and a gallon of ice cream in two days.

"Susie," Eve said to an imaginary person at the table next to ours, "what could Barry have done? How could he have used that time constructively for himself?" Turning to me and smiling warmly: "How did it make you feel to eat that stuff? I'm so glad you came to the meeting tonight, Barry. You really needed the meeting, and you felt good enough about yourself to come even though you didn't have a great week. But this coming week's going to be better."

I pushed the point: You won't talk about the conflict with the spouse at all?

"Absolutely not. We're not equipped." She paused just long enough to anticipate my next question. "If your point is that we're not dealing with the *real* problem, I say poppycock. We're dealing with the overeating problem. The marital problem is a different problem. If the fight with the partner is around eating, there are specific kinds of helpful ideas that can come from other members in the room about dealing with discouragers."

The guiding principle in such instances, Eve explained, is that people criticize you because they themselves are hurting. Quite likely, they're unhappy with their own weight. So the best response is just to let them vent those feelings without becoming defensive or angry in return.

Maybe so. But might there not be another reason for some friends and spouses to react negatively to a Weight Watchers convert? Perhaps, I suggested to Eve, they don't understand the person's new ways of thinking and behaving or they feel left out because the person has a new "family."

Eve's rejoinder was that this is just another example of how people stereotype overweight people. If they do nothing about their fat, they're considered lazy or stupid, employers discriminate against them, and they're forced to buy clothes in special stores. Yet if they slim down, some will say they've given up their real selves or become zealots. Hogwash, said Eve.

Good point, said I, finishing the last of my fish. But now that she'd brought up the matter of prejudice against the obese, another question came to mind. Doesn't Weight Watchers itself put forward a message that it's bad or pitiable to be heavy? The cofounder of Weight Watchers, Albert Lippert, has been quoted as saying: "It's simple, really. Fat people are not happy."

While this is a popular assumption about the obese, recent surveys, I indicated to Eve, suggest it is inaccurate. Overweight adults are no more anxious or depressed than others, one study found. Another found that 49 percent of women and 55 percent of men who consider themselves overweight said they feel good about their appearance nonetheless.

Eve took another tack. "We're not saying it's *bad*," she said, "we're saying it's unhealthful. There are healthful weights listed on the Metropolitan Life Insurance tables. Those are the weights at which you live optimally."

Like other body remakers I've interviewed, Eve wants to hide behind the shibboleth called Health. She claims that her program makes it possible for some people to live longer and have fewer illnesses; and this benefit far outweighs any unintended harm the program might cause those who choose not to join, or who join but do not succeed.

That argument would be more compelling were it not for the conspicuous lack of consensus among scientists on these matters. Several leading authorities contend, for instance, that half or more of the obese population would be better off fat than suffer the physiological and psychological problems that accompany their efforts to diet. And some studies indicate that the healthiest people weigh fifteen to twenty pounds more than the famous Metropolitan Life tables say they should.

Not that the notion of optimal weights is entirely lacking in merit. The numbers on the Metropolitan Life tables are certainly superior to the other leading point of reference available in our culture: namely, the ideal images in the media. Weight Watchers authorizes people to feel good about their bodies while remaining considerably heavier than the official beauties they see in the magazines and on television. And Eve's employees can lose their jobs for falling *below* the specified range for their age and height, as well as for rising above.

Still, use of the weight tables results in the exclusion of some people who may want or need the services of Weight Watchers. A member who is unable (or doesn't wish) to lose the required number of pounds by means of the program's menu prescriptions is not permitted to enter the second phase of the program, the weight-maintenance classes. Even those who function well and are happy at higher weights are not eligible for classes that could help them stay at their preferred weight, nor can they become lifetime members.

When I raised this matter, Eve found herself unable to offer an argument why such exclusion should be the case. "People who look better feel better," she said, "and people who feel better function better." I reminded her that I've interviewed people who have discovered through trial and error that they feel and function best when their weight is well above what the tables deem normal.

"To be perfectly blunt," she eventually conceded, "we wouldn't have much of a business if people were running around saying they're lifetime members of Weight Watchers when they're eighty pounds overweight."

Plastic Money

The not-so-secret truth about body remaking is that it's as much about big business as it is about improved health. Americans spend $10 billion annually on diet programs and products. Between its classes and food products, Weight Watchers alone is nearly a billion-dollar-a-year enterprise, and a large franchise like Eve's can gross $5 million.

Although Weight Watchers has 75 percent of the U.S. market for weight-loss lessons, that still leaves room for several huge competitors. For instance, The Diet Center, a franchise operation with 2,000 outlets nationally that offer individualized counseling and specialized diets in addition to group classes, racks up revenues in the $40 million range.

The national market for weight loss is truly gargantuan. According to surveys, 55 percent of women and 41 percent of men believe they're overweight, and 65 million Americans are dieting. And people who consider themselves overweight tend to be repeat customers for the diet industry. The vast majority of dieters try many programs during their lifetimes—only five to ten percent of those who lose weight manage to keep it off for more than two years. As a result, diet programs are constantly advertising in an effort to steal one another's members, and to pick up those who have given up on their current program.

Much as the existence of a massive defense industry ensures that the nation will build more and more weapons, the sheer size of the diet industry guarantees that we'll buy additional weight-loss products. Supply helps to produce demand. In both cases, the industries pump a great deal of money into research, advertising, and public relations to persuade us that by supporting them we improve our own chances of survival.

Yet the fastest-growing field in body remaking is not weight loss but cosmetic surgery. The number of cosmetic surgery operations performed in the U.S. doubled between 1981 and 1987. Today, 600,000 operations to make people look younger or more beautiful are performed annually.

Women make up the vast majority of cosmetic surgery patients, which is hardly surprising, given the differences in how men and women express vanity and the fact that breast enhancement alone accounts for close to one-fifth of all cosmetic surgeries. Were a biceps augmentation procedure perfected, we might see men swarm into cosmetic surgeons' offices.

Even so, one-third of men in a national survey (as compared to 45 percent of women) said they would consider having plastic surgery. With articles in business magazines singing the praises of plastic surgery, the male market is growing, and for some procedures it is already strong. Men account for one-quarter of all nose jobs (rhinoplasties), for example, and nearly one-fifth of eyelid surgeries (blepharoplasties).

In all, plastic surgeons rake in about $5 billion a year. Advertising budgets at large cosmetic surgery centers can exceed $1 million a year, for procedures which typically cost $1,000 to $4,000 each. In some cities, though, a nose job can run $6,000, and a face-lift can set you back $10,000.

And that's just the cosmetic surgeon's share. Spillover from the plastic surgery boom is also generating handsome incomes for other categories of body remakers who provide ancillary services. Allison, the beauty-makeover artist, relies upon referrals from cosmetic surgeons for half of her income.

Women who've had their stomachs tucked or their breasts reduced, said Allison, often notice that their faces, which are still full, no longer match the rest of their body. Their cosmetic surgeons send them to Allison's Upper East Side Manhattan studio, where they get a new makeup regimen and hairstyle.

Women who have had face-lifts also come to Allison for help. "They get their new faces, but they still wear their old makeup," she explained. "With a larger face, you can wear more makeup. The filling in of the nose, the building up of the lipline, the camouflage on the cheeks . . . all these things they did before their face-lift to create the illusion of being tighter and thinner. After the surgery, they still apply all that gunk, and it makes them look like harlots."

Increasingly, the women who come to Allison for these postoperative services are in their thirties and forties, rather than in their fifties and sixties as was the case several years ago. Nationally, over one-third of patients who have face-lifts are younger than fifty.

Even though it's good for her business, Allison disapproves of this trend. "I'm not some sort of moralist," she said. "Anything somebody wants to do to herself is okay with me. But most of these girls, all they really need is some simple beauty tips."

To illustrate, she removed from her shoulder bag a copy of a nationally famous women's magazine, turned to a layout about beauty makeovers, and gave me a concise course on how to look more beautiful. "I was the makeup artist on that shoot," she announced, "and ninety percent of the difference between the 'before' and 'after' shots was unbelievably simple. First, I got the girl to smile. A smile makes anyone instantly look ten times better. Then I had her straighten up her posture. If you stand up straight you instantly lose five pounds.

"Then I put a little concealer and some powder to get rid of the dark circles under eyes. She could have paid a plastic surgeon $4,000 to accomplish the same thing. Then I located her cheekbone. You'd be surprised how many women say to me, 'I don't *have* a cheekbone.' I say, 'Trust me, you do have a cheekbone, and you can give the illusion of having a *high* cheekbone by placing your blush under your cheekbone instead of on it or above it.'

"From there on, it's all a matter of drawing attention away from some spots and onto others. If a girl is a little overweight, the first thing I'll do is put some dynamic earrings on her to direct everything to her face."

If effecting a transformation is so simple, I responded, why are women selecting the plastic surgery options instead?

"They're lazy," said Allison. "It's easier than learning how to use makeup and accessories properly and working on your face every morning. Why does somebody go to a surgeon for a weight problem instead of eating properly and getting some exercise? It's the same thing."

Not necessarily. For many people, cosmetic surgery is enticing precisely because they think of it as an active endeavor—as one more piece in a comprehensive health-and-fitness program.

That is certainly the way cosmetic surgeons are promoting their services. An advertisement for a clinic in California features a photo of an attractive woman, captioned: "It's important for me to look and feel the best I can. That's why I eat the right foods and exercise. And that's why I had plastic surgery."

This view of cosmetic surgery stands a good chance of winning public acceptance over the next several years, given our tendency to confuse beauty with health. But some plastic surgeons are strongly critical of it. In an editorial in a professional journal, Robert Goldwyn, a surgeon from Massachusetts, warned his colleagues "of an alarming rise of hucksterism, a fearful professional pestilence." Appalled by a letter another surgeon had mailed to his patients, encouraging them to have excess fat removed and their breasts enlarged or reduced in preparation for the beach season, Goldwyn wrote sarcastically: "Our surgical Lancelot, thank God, is ready to take up scalpel and suction for the afflicted. I was not surprised that he, so busy to benefit from the beach, did not tell his patients to protect themselves from the sun."

Other doctors have raised more sober concerns about the healthfulness of operations performed primarily for reasons of vanity. Some complications of cosmetic surgery occur frequently enough, these physicians point out, that they should not be taken lightly; among the most common are scarring, secondary infections, bleeding, and skin discoloration. Less common but more frightening are nerve damage and loss of sensation or motor ability.

Public health advocates have suggested that cosmetic surgery is a drain on the nation's health-care resources. Forty percent of the board-certified physicians who are members of the American Society of Plastic and Reconstructive Surgeons restrict their practices to beautification surgery. Rarely if ever do they perform reconstructive plastic surgery such as emergency-room care, cleft-palate operations, or the repair of injured limbs.

Remember Stuart, the plastic surgeon in Atlanta? "Fourteen years of expensive medical education to take off wrinkles from somebody's face is a travesty," he railed when I raised the topic of cosmetic surgery. "In this country we are not in a situation comparable to the Third World yet; we're not bankrupt, we can afford these luxuries. Still, it is insane. Where is the justification for helping people to run away from the reality of growing old, when your time could be spent on those who are truly ill?"

Stuart's orientation to his profession is expressed in his office decoration. The only wall hangings are his diplomas; the only artworks a half-dozen clay and plastic sculptures of human hands, breasts, and noses scattered among the book-shelves. The focus in the stark white room is his paper-cluttered metal desk and opposite it two chairs intended for patients and their partners.

For our interview I sat in the chair to his left, and Stuart, a heavyset man with a bulbous nose, was seated behind the desk. "Too many accident victims and cancer patients in this city need my services for me to spend my time on vanity surgery," he said fervently.

Not that he's a purist. He does perform some cosmetic surgery, he let me know. It's just that he chooses carefully which patients to take on and restricts the cosmetic part of his practice to one-quarter of his time—"much to the chagrin of my wife," he added, "who for the life of her cannot understand why the Milquetoast dermatologist at the end of our street makes twice what I do."

How does he decide which applicants for cosmetic surgery to accept, I asked.

"If you're really aging prematurely and you have deep jowls and neck wrinkles that keep you uncomfortable, wet, and macerated in the summers, that's a valid reason," Stuart replied. "But to do it because the 'in' thing to do this year is to get a face-lift for your forty-fifth birthday, I'm not willing to participate.

"Some major corporations practically make it a condition of employment that senior people have a face-lift or some work done around their eyes. If I'm approached by a reasonable fifty-year-old executive who comes in here and says, 'Things are tight, I need to look a bit younger and less tired if I'm going to have a chance,' I say, fine. It may not be absolutely the right reason for it, but it's reasonable, it's an acceptable reason for him to be asking. Compare that to the woman who comes in saying, 'My husband's fooling around, and I've got to get him back.' I would turn her away."

Why should the businessman's request be deemed more reasonable than the wife's? I asked. Isn't that sexist?

"After the face-lift, her marriage is still going to be lousy," Stuart contended. "She'll return a month later angry with me. 'He still ran away with my neighbor, and I'm going to sue you for all you're worth.' It is your job as a physician to decide what are reasonable goals and expectations of a patient."

The principle Stuart was trying to get across—and the one that guides his medical decision-making—is that cosmetic surgery is warranted only when it will improve how a person functions in the world. To illustrate his point, he took a couple of photographs from a drawer.

The first was of a very attractive black woman. "Here's an example of a woman I did choose to work on. She was a newspaper reporter with ambitions to be in television. She was obviously quite bright—all A's as a student at Emory— and there was little doubt she could succeed in her chosen profession, except that she had exaggerated African features. In a Caucasian-oriented industry, she would not be considered attractive. I pulled her jaw back, narrowed and raised her nose, and sent her to a colleague for some orthodontic work." He explained that this woman, unlike the wife with the philandering husband, could be helped to function better by means of cosmetic surgery.

Then Stuart picked up the second photograph, but before showing it to me he asked, "How many fingers does Mickey Mouse have?" Try as I might to remember what Mickey looks like, to figure out why Stuart would be asking me this question, I could not.

"Most people don't realize that Disney characters have only three fingers and a thumb," he instructed, and showed me the picture he was holding, which was not of Mickey Mouse, but of a human hand with four fingers.

"You can do everything with four fingers you can do with five," Stuart said. "The brain switches over right away. And yet accident victims will fight and not be willing to lose even a part of a finger. They'd rather spend their whole lives babying an abnormal part that would have been better off in the bucket so they can look 'normal.' The reality is, most people they come in contact with won't even notice that they have only four fingers."

His argument was compelling, and yet I wondered how well one can distinguish between function and aesthetics in actual practice. Maybe the absence of a finger would cause people embarrassment every time they're asked to shake hands, or lose someone a high-paying sales job. Who can say in advance?

Ironically, in the case of cosmetic rather than reconstructive surgery, the line between practicality and beauty may be even harder to draw. We live in a society in which attractiveness brings rewards, and so the odds are good that an operation that makes a person more beautiful will also improve how he or she gets along in the world.

Patients who undergo cosmetic surgery are usually pleased with the results, according to studies, and are perceived by others as more sexually appealing and responsive and as better marriage partners after their operations. One researcher reported improvements in the personalities of fully two-thirds of a sample of patients following cosmetic surgery. Contrary to Stuart's expectations, studies have found that lifting a woman's face or breasts will often boost her spirits and her marriage.

Determining what constitutes an improvement in functioning can be difficult, as I tried to suggest to Stuart by way of the example of a young woman I interviewed. She did well in school and had plenty of friends, but she was preoccupied by the fact that her breasts were larger than any of her friends'.

They were not so large that they caused back problems, however, nor did they create difficulties in buying clothes. For those reasons, her family physician refused to give her a referral when she decided to have a breast-reduction operation performed.

By her own account, not much changed in her life after she had the operation, except that she felt less self-conscious, and men on the street stared at her less frequently.

In this case, like many in cosmetic surgery, little improvement in functioning was gained. On the other hand, does this distinguish cosmetic surgery from every other medical specialty? Some allergists make a good chunk of their incomes by administering weekly shots to people whose allergies are not severe enough to handicap them, nor even to cause discomfort for more than a few weeks every year. Often the treatments do not help these people anyway.

"But let's carry your analogy one step further," said Stuart. "Granted, people have a natural urge to look normal, just as they do to sneeze in the presence of allergens. But it's getting to the point that we are spreading the infection we're charged with treating.

"How would it be if allergists placed scratch-and-sniff advertisements in magazines, advertisements that contained chemicals that caused readers to react? That is essentially what cosmetic surgeons do.

"Plastic surgeons shouldn't promote the disabling notion that beauty is a valuable end in its own right, any more than oncologists should promote cigarette smoking. I've heard intelligent people come up with good arguments for that one too, by the way. 'Smoking reduces stress.' 'If it helps them keep their weight down, maybe it's not so bad.'" Stuart quoted these comments with a pronounced sneer, but then a sad expression spread across his face.

"I'll tell you a perversely humorous story about the clinical dangers of vanity," he continued. "A woman I operated on last week is the mistress of a local physician, an internist. She had a ten-centimeter tumor in her breast. Malignant. He claims he didn't notice it.

"How could he not notice it? I could believe that maybe someone, even a doctor, would overlook a tumor in his *wife's* breast, but how he could fail to notice one in his *mistress's* breast is beyond me. I guess he wasn't thinking of her breasts in any manner except as objects for his sexual pleasure.

"But then," he added, "she wasn't much better in that regard herself. The day she came to my office, she was wearing a low-neck dress and had put makeup on her cleavage to hide the age lines. This was after an oncologist had found nineteen positive nodes under her arm, in addition to those in her breast, and told her she'd have to have major surgery."

Playboy as a Tax Deduction

Stuart's receptionist interrupted us at this point with the news that we'd gone overtime and patients were waiting. He had one more point he wanted to make, though, before I left, and to do so he took out another visual aid from his desk. It was the current issue of *Playboy*, which seemed out of place in this austere office.

"The Saks Fifth Avenue catalog of our profession," he said, holding the magazine in front of him. "If you want to know which breast shape is stylish this year, just open up and look. Want to determine whether we're going with Irish turned-up noses this season or the Swedish bob?"

There was anger in Stuart's voice.

"It is travesty enough that our profession actively promotes vanity. On top of that, we practice inferior aesthetics. A woman can walk into a cosmetic surgeon's office—a thin woman of Mediterranean ancestry, let us say—and ask for a northern European nose and thirty-eight-inch breasts. If her credit rating checks out, the surgeon will comply with her request, even though her face will look asinine as a result, and her figure will be all out of proportion.

"There are objective procedures one can use in plastic surgery. With a rhinoplasty, for instance, you measure the vertical height of the face, how much projection the chin has, the space between the eyes, the relative location of the cheekbone. Then you build a nose that will blend in with the other features.

"For a patient who seeks a simple rhinoplasty to remove a bump or straighten out a crooked nose, these kinds of considerations are all that come into play. They should be in every other case as well, but try to tell that to a Jewish mother who brings you her daughter and a photograph of Brooke Shields and demands that you fit the kid with a schnozz like the one in the picture."

He thumped the copy of *Playboy* on his desktop. "The IRS has ruled that a subscription to *Playboy* magazine is a tax-deductible expense for a plastic surgeon. And frankly, I cannot honestly dispute the ruling. It's a tool of the trade in some practices."

Stuart's comments point up the extent to which bodies themselves have become objects to be sold in American society. Surgeons sell not just corrections to the body—to make us look more normal or attractive—but something far more transitory, body *fashions*. They alter the size and shape of our buttocks, breasts, noses, or eyes to fit current styles.

If the period from the mid-1800s through the mid-1900s was one in which the individual's body became a vehicle for the display of consumer products, lately we're faced with the next logical step in consumerism. No longer can we merely dress up the body we happen to have, or improve it by losing weight or having a beauty makeover or straightening out the curve in our nose. We must actually purchase a "new body."

In a literal sense, of course, it's impossible to trade in our bodies for new ones. Even if a plastic surgeon altered every region of our anatomy, the best we'd end up with is a revamped version of the body we were born with. The crucial point is rather that so many people *think* in terms of getting a new body that it makes sense these days to *talk* in those terms. Where else but in modern-day America could a magazine titled *New Body* find an audience?

The mounting enthusiasm for cosmetic surgery actually represents only the most glaring evidence for this latest view of the body. The notion of gaining a new body is most highly developed elsewhere—within the contemporary fitness movement.

If cosmetic surgeons offer to change our looks and, as a consequence, our prospects for love or wealth, those who promote exercise hold out all that and more. They say the new body which results from proper exercise will keep us alive longer, save money for our employers, and raise our social status.

TEN

Health Club Hawkers

The idea that a person's body is not a given but can be transformed dates back to Europe in the 1600s. Prior to that time, a man was judged suitable for the military on the basis of his God-given shape. To become a soldier required broad shoulders and big thighs. But as French theorist Michel Foucault pointed out, during the seventeenth and eighteenth centuries a soldier came to be seen as an object that could be molded, as if out of clay. A recruit's posture was corrected until he learned to stand stiffly and throw his chest out and shoulders back, to fix his eyes and stare directly in front of him, to remain motionless unless ordered to move. By way of this training, most of it focused on the body, common peasants were converted into infantrymen.

Over the intervening centuries, more and more people— from shop clerks to executive trainees to television newscasters—found themselves called upon to alter their bodies for the sake of their social status or their careers. Now, late in the twentieth century, millions of civilians of both sexes voluntarily subject themselves to rigorous physical discipline—at health clubs.

The *image* health clubs project is not one of drill sergeants and Spartan living, of course, but of satisfied customers in luxurious surroundings. Most people think of health clubs as high-glitter establishments like the Century West Club or Sports Club LA, or the Vertical Club or Excelsior in New York—places that make it into movies and magazine articles and can cost thousands of dollars a year.

That image is inaccurate. Most of the $4 billion health club industry is located not in high-rise palaces overlooking movie lots or the Manhattan skyline, but in the American heartland, in the 7,000 or so body shops scattered around the country which charge only one to several hundred dollars each for an annual membership.

These clubs employ armies of recruiters who are exercise buffs themselves and whose job it is to spread optimistic messages about the joys of exercise.

"I Love What I See Happening in This Country"

"Americans are finally starting to realize that we all have control over our destinies if we take it," said Sammy, who has worked for a number of health clubs and who hosted a televised exercise show five days a week from 1961 until 1985 in the Midwest.

"It's mind over matter," Sammy told me during our interview at his suburban ranch home. "If you're overweight, you exercise and change your eating habits, and you accomplish what you're after in life.

"Everybody should exercise. You feel good, you feel exuberant when you get out and exercise. Exercise is like an insurance policy. If something happens to you—you get into an accident or get some kind of disease you have no control over—if you're in shape, your chances of recovery are better."

Sammy was animated as he spoke. This is a message he

deeply believes, and one he's delivered tens of thousands of times, maybe more, to every interviewer or civic group that will listen.

"I'll be honest with you, Barry," he said, inserting my name Dale Carnegie style, although we'd just met. "I was running when it wasn't fashionable to run. I would go out to the high school track when they first built it in 1958. I remember running the streets in a white pair of shorts and a white T-shirt. People must've thought, 'This old guy running around in his underwear, throw a net over him. He must be some kind of nut.' "

At age seventy-three, Sammy still runs twenty miles a week.

He also swims occasionally, pumps iron in the gym in his basement, and logged 7,000 bike miles in 1987 alone. The day before our meeting, Sammy rode sixty miles with a buddy half his age to visit a friend out in the country.

"I love what I see happening in this country, Barry," he continued. "People are doing something for themselves. They just feel they got a better shot at keeping their youth. And it's good to stay young. A guy gets to be forty, the companies don't want to hire him anymore. You go into a large corporation, you go to GE, IBM, any of those big companies, and you take two men who are of equal ability and seniority, and leaving the politics aside, they'll take the guy who's physically fit over the guy who's not, and promote him, for the simple reason that they're going to get more time out of him. The fellow who's not in shape is going to be off five or six weeks a year because he catches everything that comes along. I just read somewhere that a physically fit executive earns $3,000 more per year than the guy who's out of shape."

Sammy, who was dressed in a red-and-white warm-up suit, white athletic socks, and white Nikes, related all this in the first five minutes after I walked in the door—a door that sports not one but two "Thank You for Not Smoking" signs.

We sat in his paneled living room/den, and for a good hour he effused about the practical and spiritual rewards of having a well-exercised body. When he finally came up for air, I managed to shift the discussion to his own biography. He'd been athletic since childhood, he told me, and a salesman since adolescence, but not until well into midlife did he make his living selling athleticism.

He served in the Air Force during World War II and, after his discharge in 1947, peddled household goods on the installment plan, door to door. During his spare hours Sammy worked out and played football on amateur teams. The idea for his TV exercise show came from a discussion at the Y in 1960 with a fellow who was working at a new local station. Like many at the time, the station was hungry for cheap local programming.

Very likely his was the first exercise show in America, but even in its best years Sammy couldn't support his wife and two children without second and third jobs to supplement his income. He'd leave the studio each morning and drive to a church that let him use a large hall. There he conducted live calisthenics classics for small groups of housewives who paid a couple of dollars apiece. Afterward he sold them vitamins he packaged himself, with his picture on the outside wrapper. Most years he also worked part-time in a used-car lot owned by one of his friends.

During the twenty-five years the show was on the air, Sammy's half-hour time slot was shifted back and forth between mid-morning and sunrise. Every time the show built up viewers and advertisers, it was moved. What finally brought it down was the mammoth exercise industry of the 1980s. As Sammy said: "With VCRs, you can buy your tape and keep it, and then you can exercise any time of the day you want. And it's getting to where there's a health spa or aerobic exercise class on every corner."

He tries to be as upbeat about these developments as he is about everything else related to fitness, but the fact is that he has serious reservations, and some bitterness. "They give you only one 'right' way to do it," he said of the videos, "and people come away feeling, 'My God, I can't do this damn thing,' and they give up on exercise. When I had my program, if I was doing a leg lift I'd show you how to do it with your legs straight, toes pointed, knees locked. But I'd tell you, 'Susie, if it's a little difficult for you, bend your knees slightly. If that doesn't feel right, Nancy, raise one leg at a time.' And I kept yelling all the time: 'Okay, beginners, drop out! Beginners drop out now.'

"Some people wouldn't listen. I got thousands of letters from husbands. 'You're ruining my sex life, my wife's too stiff to do it.' "

In spite of some efforts a few years ago to raise money to make his own video, which he planned to wholesale to local video shops, Sammy hasn't been able to cash in on that part of the exercise market.

Health clubs are another matter. Since the 1950s, the decade of "figure salons," Sammy has been sought after by owners of health clubs looking for a high-profile instructor or recruiter. Not until the proliferation of clubs during the 1970s did any offer him enough money to make it worth his while.

He enjoyed the work when he first joined up, but, he confessed, he became dismayed after a few years. "The rule is: Get the money in and load the place up. You come, there's basically nobody there to give you any real help. If you come after work, you're lucky if you can get a parking place. It's so packed you can't get to the equipment, you have to stand in line.

"When these places first started opening up—I'm talking, what, fifteen years ago now—they had instructors there to program you and work with you. We wanted a satisfied customer. We would nurse you. 'Barry, how you doing, here's a new exercise for you,' and 'Hey, you gained a pound, let's work it off.' So you figured, 'Jeez, this guy cares about me.' It's bad they don't do that so much today, because it's so easy to give up something. People will start any kind of plan that will make them feel better about themselves, get all enthused about it, and then give it up."

Sammy's expression was one of exasperation as he spoke of these clubs, his disappointment that of the small businessman whose personal rewards derive as much from pride in the quality and integrity of his product as from the profit he makes, and whose industry has been overtaken by conglomerates.

"I wasn't fired, I quit," he made sure to tell me. "Anyone I signed up, I made damn sure they got service. But you can't do that with the volume they demand. I wrote up a million dollars of business in three years. The outfit I worked for had eighteen spas in the region, and the girl at the central office would call all eighteen, four times a day: 'How many appointments you got set up? How many have you written up?' "

Another reason he quit, he explained, is that the club interfered with his own fitness program. "I didn't ride a bicycle the whole five years I worked there. Didn't have the time. I was in there six, seven days a week. Weekdays, I'd get off the air at six-thirty in the morning, shower and eat something at home, and be at the spa by seven forty-five for the before-work crowd. Sometimes I'd be there until nine at night."

What It Costs to Belong

Since 1985, Sammy has been in what he calls "active retirement"; his time is spent biking and giving rousing lectures to high school students and civic clubs, and he hasn't set foot in a health club. The complaints he voices continue to be heard throughout the country, however. Consumer groups and Better Business Bureaus regularly receive complaints about high-pressure sales tactics, crowded conditions, and misleading offers.

Another problem is the large number of clubs that close every year, leaving those who paid for multi-month or "lifetime" memberships out in the cold in their sweatpants. A tally by the New York State Attorney General's Office showed that in a recent two-year period nine clubs, with a total membership of 3,600, closed in the Albany, New York, area alone. Sometimes clubs look perfectly healthy before they close, and they sign up long-term memberships at the front counter while bankruptcy papers are being prepared in the back office.

In the late seventies, the Federal Trade Commission conducted an extensive investigation of health clubs throughout the United States. The FTC documented widespread overcrowding, injuries ranging from muscle strains to heart attacks and strokes, and cancellation policies that made it nearly impossible for members to get refunds if they moved or became ill. A 200-page report concluded that the health club industry needed to be regulated.

During the early 1980s, it looked as if the FTC might pass a rule that would limit memberships to a maximum of two years and force clubs to provide prorated refunds to consumers who dropped out. Instead, it announced in 1986 that the proposed regulation would handicap the industry, and that state laws could take care of the problems.

Most states have in fact enacted legislation to protect consumers, but only a few of these ordinances have much muscle. Two that the National Association of Attorneys General considers especially effective are those of Connecticut and Maryland. Connecticut requires health clubs to contribute to a fund that the state uses to reimburse consumers when clubs close. Maryland charges every club a registration fee, which goes to pay the expenses of an enforcement group that oversees operations of the health clubs throughout the state.

No amount of legislation can protect us, though, from the health club industry's most powerful weapon—membership recruiters who know how to poke at our insecurities until we sign up for the new body they claim they can provide.

The nation's health clubs employ hundreds of thousands of men and women who lie in wait, Special Introductory Offers in hand, for those of us who feel overweight or undermuscled. Few of the recruiters look like Sammy. Most are young and beautiful, like a woman I interviewed in Richmond, Virginia.

Kristin, a shapely twenty-year-old khaki-blonde in a bright red tank top and tied-at-the-hips black skirt, had been in the exercise business only a year but was already a pro salesperson.

Our question-and-answer session took place at ten o'clock one night in February at a Burger King down the street from the health club where she was employed.

"I used to work at a tanning salon in the shopping mall," she began. "I got free tanning sessions when we weren't busy, but it was totally boring, and you couldn't make any money. I'm making like five times as much at the club."

The only thing she misses from her old job, she said, is her tan. Her current job came about by chance. "I was standing by the checkout counter at a restaurant applying for a waitress job," Kristin recalled. "These two guys came in to pick up their carryout order. I guess they noticed me or something. So they walked over and told me they were managers at such-and-such health club, and would I come down there and fill out an application to work for them and see how I liked the place. They explained it all to me, the wages and commissions and everything."

She took a long sip from her Diet Coke and ate a couple of french fries. "I think they chose me pretty much on looks. If I hadn't been as well presented as I was that day, they probably would have looked right past me. I was wearing a short skirt and nice top, and they told me I looked like I was in good shape."

I asked if they were right, if she had in fact been exercising.

"Nope. I hardly even knew what a health club was. I didn't work out or anything. I'd played volleyball in high school, and I used to swim in the summer a lot. But I wasn't doing anything then, I was a total slug.

"When I first started at the club they told us that all new employees had to work out three times a week. With the training for the job—they trained us every single day for seven weeks, and I was still working part-time at the tanning place—I hardly ever got in there for my workouts. The first time I ever really worked out was this last month."

Kristin said she'd become interested in free weights, and flexed her left arm to show me her biceps. But she explained that she doesn't wish to change her body very radically.

"I don't need to work out for my weight or anything, because I'm naturally thin. I never had any problem with being flabby anywhere. A lot of the girls who work at the club are short, and they have to work off their excess weight."

I asked if most of the recruiters at the spa were her age. "Oh yeah," she responded, "they hire young girls so they can bring the crowds in to join. The whole idea is to create an atmosphere where guys talk about the place—'You should see the girls who work there,' that kind of thing.

"Matter of fact, there's this thing we call the 'sex closing.' That's where you tell a guy you're going to go on a date with him, to get him to sign the membership paper. Then, if he joins and comes to collect on the date, you just blow him off. Girls do that with guys all the time anyhow. Tell them you got a boyfriend. Guys don't really care, because once they get into the club there are lots of other girls, so the main focus isn't on you anymore."

Lest I get the wrong impression, Kristin immediately added that she had deployed the 'sex closing' only once, and that male as well as female recruiters use this tactic.

"Mostly we use what they call a buddy system," she said. "You constantly chat up the members when they're at the club and get the names of their friends and the people they work with. You call them and invite them out for a drink or down to visit the club. A lot of times I'll make the rounds to bars and give guest passes to the bartenders and waitresses to hand out. My name's on the pass, so they'll call me for their free visit.

"Plus, we put contest boxes all around the city in shops: 'Win a Free Membership' things. Everybody who fills out one of those wins a free week, or they get three weeks for five dollars. That gives us a chance to pitch them from inside the club."

That pitch is all about becoming a new person. What most people want to alter when they think about joining a health club is their orientation to life, Kristin told me. "They want to make themselves look better so they can have a better attitude. If you're not happy about the way you look, it's hard to be happy in other ways. Your whole attitude on life is just *blah*."

What about her, I inquired. Her body seemed to match the current ideal, and she'd already told me that she didn't need the services she sells. Was she one of those people for whom I'd searched so hard over the past two years—someone happy with her body?

No such luck. "I go through my stages of 'I wish I could change this or do that,'" Kristin admitted. "Like I wish I had had braces when I was a kid, so my teeth would be perfect." She smiled the full width of her face, as if performing a pantomimist's exercise, to reveal some incisors at a slight angle. "And I wish my skin didn't break out. I wish I didn't have to wear makeup and spend so much time on my hair every day. It takes me an hour and a half every morning to get showered and made up.

"If I could go to the tanning center again, I could wear a lot less makeup. When I have a tan, my face clears right up, and it just brings my features out. But people at work would get on me if I went back there."

Why would it bother them? I asked.

"Because it's fake. It's artificial. It's not natural. You know, at the club we promote the natural approach. We sell a line of natural vitamin supplements and stuff."

Is that a big drawing card for the club?

"Not really," said Kristin, who was momentarily distracted by her french fries, six or seven of which she ate in rapid succession.

"Once we get them in the club for a visit, we dig, dig, dig. We try to find out as much about them as possible. What are they after? Do they want to meet people? If it's a woman, does she want to lose weight in a particular spot, or does she want all-over toning? Men either want to build up or else lose their spare tire. Whatever it is, you give them a little taste of what the club has to offer for what they want to accomplish.

"You have them try out one of the computerized treadmills or rowers, because they're fun the first time you get on one and you can do them in your work clothes. You let them peek through the window into the sauna. With a middle-aged woman who's never worked out before, that's where you really sell the membership. Probably she'll never use much of the exercise equipment; she'll just hang out with the other ladies in the sauna."

Kristin proceeded to tell me about the various facilities at the club—the pool, the five different types of toning machines, the graded running track, the snack and juice bar—and how she's able to find some gadget or service to entice anyone who walks through the door. One selling point she failed to mention, though: the health benefits of exercise.

Perhaps she doesn't need to. Perhaps the public is so convinced of the health benefits of exercise by now that it would be redundant to advertise them. Or perhaps health clubs have devised ways to build the health message into their song and dance without making a special point of it.

Kristin finally did get around to mentioning that her club employs an exercise physiologist. "He has a master's degree," she said in a hesitant tone suggesting she wasn't quite sure what a master's in exercise physiology really is. "Once we pitch the clients, the exercise physiologist puts them through our 'Body Info System.' That's a computer program. I don't understand how it works exactly, but it gives you a personalized analysis of your body's needs.

"First he takes your weight and measurements, your heart rate and blood pressure, and he checks your percentage of body fat, using calipers. Then he sends you around on twelve different machines to find out where your strengths and weaknesses are. A lot of people, halfway through the tests, will go, 'Oh my God, I didn't realize I'd gotten so out of shape.' If that happens, the physiologist tells me about it, and I know I'll be able to talk them up to a higher level of membership.

"Anyway, he puts all the information into the computer, and a piece of paper comes out that gives a three-week program of exercises on our exclusive machines. It tells how many repetitions and the right weights and everything.

"It really works, too. If you come in four times a week and go on all the machines and do all your repetitions, you'll really see a change in your body in three weeks' time."

Kristin, accustomed as she is to paying attention to a prospect's nonverbal cues, noticed the look of incredulity on my face.

"For sure," she reassured. "You see a difference really fast in somebody who's out of shape. The hard part for people is after the first couple of months. They look better, they feel better, and things kind of level off, where you don't get a whole lot more improvement real fast. Then they have to rely on their own willpower to keep at it.

"To get *totally* in shape takes a year, which is really worth it, because after that, all you have to do is just maintain what you've done."

How, I asked Kristin, does she handle prospective members who hear the sales pitch but still aren't convinced?

"Usually they don't want to spend the money. They'll say they can't afford it. So I'll go after their weak spot. Like if I know they smoke, I'll go: 'Stop smoking and put that dollar toward a health club.' Or if they're overweight: 'Give up two Cokes a week and you'll look great.' A lot of times women will say their husbands don't want them to spend the money. So I'll go: 'Wouldn't he want you to lose some weight?'

"You have to tailor what you say. You try to be really respectful if it's an older person, where with a younger person you have to be assertive. 'Listen, you told me you wanted to work out, so stop fooling around and get your butt in gear. You know you're going to be happy that you did it.'"

Kristin said this in the tone of voice of a girl trying to bully her older sister into loaning her a sweater.

Then her voice took on an air of quiet confidence: "With a guy, it's easy. Every guy is insecure about his body. All I have to do is find his weak spot. 'We need to do a little work on your upper body, don't we?' Or, 'You're going to be skinny your whole life if you don't get in here and do something about it.'"

I was convinced that if she set her mind to it she could get anyone to sign on the dotted line, and I had only one more question for Kristin. So far, who had been the toughest person to sell?

"My mom," she said without a moment's hesitation. "When I started at the club, I tried to get her to join. A lot of the girls sell memberships to their families, because it's an easy way to make your quotas when you first start out. I'm really close to my mom, she's a great mom, but there was no way. . . . She's naturally thin like me; people say we look alike. She has a really cute figure for somebody who's about to turn forty-five. There's no way she'd join a spa.

"Besides, she's totally busy right now. She went back to get her college degree, and she's in all kinds of clubs at her church. She wouldn't use her membership."

Pubs and Town Squares

Whatever else may be true about it, the modern health club is an institution that keeps people occupied. From a sociologist's perspective, a "fitness center" is a building full of men and women who might otherwise stay home and fight with their families, or go out drinking and driving, or just remain bored and disgruntled. A prime social function served by the health club industry is to direct such people into places where, instead of being a drain on society, they build up their strength, their egos, and their local economy.

Health clubs serve to track people in another way as well. They provide employment for a group whom our society is committed to keeping happy but for whom there are not enough adequate jobs—the well-behaved, entirely average children of the working and middle classes. Exercise emporiums are among the few places in our high-tech society where someone like Kristin, who attends a state college part-time but isn't interested in school, can land a relatively safe and glamorous job.

Unlike the body-remaking professions, the health club industry allows a person to make a career without special credentials or technical skills. This is true not only for those who sell memberships but also for the men in fancy gym shorts and women in tight leotards who lead aerobic workouts and show members how to position their bodies on the machines.

A recent survey conducted by the Aerobics and Fitness Association of America revealed that, of the 100,000 or so health club instructors in America, only one in ten has had formal training. This number is likely to increase rapidly, though, because at least fifty certification programs have sprung up.

As in many other fields, these programs vary considerably in their levels of sophistication. To obtain an instructor's certificate from one place in Colorado, you need do nothing more strenuous than send $75, along with a videotape of yourself leading a class and a statement that you've had 100 hours experience and taken an exercise class.

In contrast, a training center in Dallas requires two courses, which together cost $1,000 and involve 80 hours of training in everything from how to perform specific exercises safely to lectures on anatomy, physiology, and nutrition. Students must pass written and performance exams in order to graduate.

But what seems to count most toward success in the fitness business—aside from good looks—is an understanding of human motivation. And while the more extensive certification courses for health club employees do offer instruction on how to motivate, the vast majority of health club workers rely on their own intuitive abilities.

Bill, who at age thirty-three owns a half million dollars' worth of shares in local spas throughout the Northeast, is an example of how well one can do in this business on pluck alone. He consults with several small health club chains (of two to ten shops each) that pay him big bucks to help with their advertising and promotion. Bill currently lives on Long Island and commutes an average of 1,000 miles a week in his new Audi 5000, though just six years ago he was working on the assembly line at a Ford Motor plant in Pennsylvania. A gym rat himself, Bill found work in the weight room at a health club when he was laid off at the factory. After a few months there, he persuaded the owner to let him branch out and help with other parts of the operation.

"I basically saved his business," Bill said during an interview he squeezed into his schedule by taking me along for the two-hour drive between a couple of his accounts. "I snooped around all the clubs in town and figured out what made them tick. Most guys in this business don't use their heads."

He brought ideas back to his boss, who eventually let him write radio and newspaper advertisements which, Bill claims, doubled the membership roll within a year, despite competition from three new clubs in the area. He said he's used the same basic principles in his work ever since.

Bill's first rule: Avoid pictures of nearly nude nineteen-year-olds with perfect bodies. Cheesecake and beefcake may grab a prospect's attention, but those aren't the new bodies that people in mid-America are after.

"Management always wants to use those pictures, because it's their idea of the club's image," he reported, "or the manager is dealing with his own ego. It's a turnoff to the actual customer, unless your club's out in Hollywood somewhere. Believe me, I make it my business to listen to people about these things. I take members from our clubs and I show them the competition's advertising. I ask what they think. And people aren't fools. They know they're never going to look like Cher. Most of the towns I work are pretty conservative anyway. Women say to me: 'The girls in that picture look like whores.'

"If you're joining a club you want to know what's in it for you. Period. If you're taking up racquetball you want to know what kind of training you'll receive or who's there to play racquetball with. If you want machines, you want to know whether you're getting basic Nautilus or some of the new fancy equipment and whether there's an instructor who can help you build your program. If it's aerobic dance, you want to know what you'll get here that's more sophisticated than at the local church that offers it for two bucks a session."

Rule number two: Hit 'em when they're down. "Everybody and his brother knows that people buy memberships after Christmas because they got fat over the holidays, and in the spring so they can fit into their bathing suits.

"I happen to believe you can target your pitch every day of the year. I run my radio and TV spots for housewives at the mid-morning low. Our message to Mrs. Joe Q. Public is to get out of the house and rejuvenate. With men, I won't advertise on radio except between five and six in the afternoon, when they're tired of driving the same route home every day and don't want to face the wife and screaming kids. They're feeling the tension in the back of their neck, and they know from the magazines that working out will help."

Though Bill sounds like a huckster, I was struck by how much he thinks like an anthropologist.

"During the last twelve months, I took my first two vacations ever," he said out of the blue as we drove along the Pennsylvania countryside. "I'm a seven-day-a-week, fifty-two-week-a-year kind of guy. But my wife was really pushing me. So I blocked out a week and told her I'd go anyplace where the dollar was strong and the flight took less than half a day.

"She chose Mexico. And you know something? Every little village you go to down there they have a town square, and every night the whole town is in that square. The men are on one side and the women are on the other. Young couples are walking arm in arm.

"Well, it occurred to me," he went on, "in our country that's what a health club is. I don't care what anybody tells you—a health club is, *número uno*, a meeting place. That's why you're seeing so many clubs add beauty shops and restaurants to their facilities. Even the grittiest gym that caters exclusively to hard-core muscle builders has its social side.

"In a lot of the clubs I service, a primary motivation for a person to come is to see friends or make new ones. A good operator doesn't resist the social side, he caters to it."

Then Bill told me about his second vacation ever. Again, he made an analogy to his profession.

"A few of the club managers in one of my towns got together and gave me a pair of round-trip tickets to London as a thank you for a big promotion I ran for them. And every neighborhood you go to over there—doesn't matter if it's in the middle of London or out in the sticks—it'll have its own pub. They call them locals. Everybody has a local, and if you don't live in the neighborhood, or if you're dressed better than everybody else at that particular pub, you're not welcome there.

"I make fitness centers into 'locals,' " he proclaimed with a look of satisfaction. Then he reached for a more American image: "If a mom-and-pop operation is going to make it these days, when you're competing against the McDonald's and the Sears, Roebucks of the fitness industry—I mean franchise operations with two or three hundred clubs each—it damn well *better* cater to its market."

He invests his own money only in facilities that appeal to specialized markets. A mistake many owners make, he explained, is to try to be everything to everybody, thereby turning out to be nothing—or nothing special—to any particular segment of the market.

By way of example, Bill mentioned someone he worked with a few years ago who owned a club that was doing poorly in a city of 400,000. "The guy had three things going for him—location, location, and location; only he didn't know it. His operation was on the outskirts of town, right next to the big highway. It had been a warehouse for a supermarket chain before he bought it, and the only thing against it was that it wasn't exactly scenic around there. There were a couple of ramshackle old houses and a junkyard.

"But it was on the *east* side of town," he said, pausing for dramatic effect. "One thing you learn in working different cities is that rich people live to the east of town. That way they never have to drive into the sun going into work or coming home. This club was two minutes off an exit everybody had to pass on the drive home to the suburbs. First thing I had him do was put up an attractive sign in the parking lot to catch your attention from the highway. I also had him put some new siding on the building so it didn't look like a tin can. He'd already bought all the right machines, and inside the place looked pretty good.

"Once we took care of the exterior, we were home free. There was one other club on the east side, but it was part of a large chain and had no class at all. I made our facility the social center of that city.

"I call it 'our club' because I own twenty percent. I told the owner I'd put up the cash for a landscape architect and the first month of advertising in exchange for part of the business.

"For our grand reopening," he continued enthusiastically, "I sent out engraved invitations to all the business and community leaders inviting them and their wives. In the advertising I billed the place as somewhere you could sweat next to all the right people. The image slogan was 'You Deserve the Best,' and the copy was about all you've accomplished in life, and the idea that you give your best and now you should give yourself the best in return. I really made it into the Rolex of health clubs."

Bill listed the important people, from the mayor and TV anchorwoman on down, who showed up for the opening. He went on to say—as we arrived at the town where he'd be working the rest of the day, and he dropped me off at the bus station—that the success of that facility has allowed its owner to open three more clubs in the region.

Fiscal Fitness

The health club has become a fixture of the American landscape to an extent that none of the other body-remaking industries has so far achieved. Bill works with clubs located in suburban shopping malls, decaying downtowns, and every district in between; and to ensure that no resident overlooks them, he advertises their existence on billboards and city buses.

Tourists from overseas, seeing the United States for the first time, must think us the best-exercised country in the history of the world. In truth, only about 8 percent of Americans get as much exercise as the Public Health Service recommends. Just as they drop in and out of weight-loss programs, many people join health clubs, but relatively few attend regularly and many stop going altogether. In some parts of the country, only about one-fifth of those who join health clubs renew their memberships, and even the most successful of clubs can keep no more than about two-thirds of their members.

This may eventually change. In the past decade, thousands of hospitals and corporations have built exercise facilities on their property, thereby sending a message to the American public that two of the most important institutions in our society believe that attending a health club is a smart thing to do.

For both hospitals and corporations, the impetus for getting into the health club business was, predictably enough, financial.

When hospital occupancy rates hit a quarter-century low in the mid-1980s, administrators began to seek new ways to compete for patients. One has been the creation of special facilities, such as women's health clinics, weight-loss programs, and fitness centers—either on the hospital grounds or through satellite facilities. About half the nation's hospitals now maintain "wellness" programs of one sort or another, many of them devoted primarily to exercise.

Health-promotion operations can be profitable to the hospital even if they do not make much money in their own right, because they cast the hospital in a favorable light. People who have enjoyed the hospital's health club are likely to choose to have their babies and their hernia operations there, and to recommend the hospital to friends.

As a result, hospitals spend large sums of money on fitness facilities, even in fairly small markets. Take Fort Smith, Arkansas (population 75,000), for example. A $3.45 million, 43,000-square-foot fitness and sports medicine center is due to open at a hospital there in mid-1988. The building will include a circuit-training area, a three-lane jogging track, a six-lane pool, two aerobics rooms, and dining and demonstration areas.

Given that many people exercise these days specifically because they hope to *avoid* hospitals, it's ironic that hospitals have gone into the fitness business. In so doing, they certify their commitment to "preventive medicine" (the buzz phrase in health circles at present); and for those citizens who've been made paranoid by news accounts about the hazards of vigorous exercise, the hospital health club provides a place to work out where medical help is available should it be needed.

The marriage is not a perfect one, however. In an article in a trade magazine aimed at health marketing professionals, Alan Schwartz, an industry consultant, tells the story of a hospital exercise facility that failed because it was too much like a hospital. "The signage was hospital signage," he reports, "the flooring was hospital tile, when you signed in it was like checking into a hospital. There was a lot of sterility built into that facility, and that just won't work. You have to have a fun atmosphere."

It's also important, warns Schwartz, not to pollute the place with lots of sick people. Health clubs are for happy, healthy people. "I've seen clubs," he laments, "where hospitals have sent the walking wounded—people who have to be monitored as they run around the track or physically lowered into the pool. Unless these patients' hours are restricted, the hospital does the club a disservice; it has a dampening effect on other members."

If the message comes out mixed when a hospital opens a health club, no such problem exists when a corporation does so for its employees. Businesses have nothing to gain from sick employees and much to lose. General Motors once estimated that every car the company sells costs consumers an extra $175, thanks to the expense of employee medical benefits. All told, American industry pays $150 billion a year in health-care expenses.

Although some firms (National Cash Register and Eastman Kodak, for example) offered fitness programs as long ago as the late 1800s, not until the 1970s, when the costs of health benefits had started to double every few years, did large numbers of corporations follow suit. But today, at least 50,000 manufacturing and service companies offer fitness programs for their employees, and 20,000 or more have installed clubs on their grounds. Some, such as Xerox, PepsiCo, and Campbell Soup, spent millions to build large, state-of-the-art facilities. And it's estimated that within a few years one-fourth of all U.S. corporations will provide some kind of fitness program.

The corporate fitness movement illustrates a basic truth about the nature of health beliefs. Ideas on how to care for the body reflect the vested interests of those who espouse them. Traditionally in America, exercise was not something corporations sought to promote. In fact, in the early years of American industry, industrialists and Protestant clergy lectured against any leisure activity that appeared playful or might take time or energy away from labor, which they held to be the ultimate source of human satisfaction.

Nowadays, corporate directors view exercise in just the opposite way: as a tool for getting more and better work out of employees.

"Why are so many corporations investing in fitness programs for their employees?" asks a brochure sent to executives by a company that sells products to corporate fitness facilities. "The answer is simple: it pays in many ways. Studies indicate (and companies with their own programs agree) that healthier, more fit employees work more efficiently, are absent less, have better morale, are more productive, and generally feel better about themselves and their employers."

A rather beamish picture that, and probably no nearer the truth about the relationship between exercise and work than the earlier, negative view. People who feel good probably do tend to make better workers, but a corporate fitness program cannot be counted upon to manufacture them. Those who join such programs, studies have found, tend to be the employees who need it least: they are wealthier, healthier, and more involved in exercise and preventive-health measures in the first place.

It pays to read evaluation studies of corporate fitness programs with a critical eye. Some do seem to show dramatic benefits. For example, in 1983 Tenneco evaluated the effects of its health club after its first year of operation. It was found that the company was spending less than half as much on health-care costs on those who exercised as on their sedentary colleagues. Exercisers also took fewer sick days than did non-exercisers.

Neither of these facts suggests, however, that Tenneco saved money by building the health club. The authors of the study indicated that overall absenteeism rates did not change at the firm after the facility was built and that the differences in health-care costs between exercisers and non-exercisers were attributable not to exercise but to differences in personality between the two groups.

Nationally, reports vary as to the success of exercise and other health-promotion efforts within corporations. Some companies claim great savings, others do not; and according to some, the major results are short-term. When a program first starts, employees enthusiastically join up and absenteeism drops, but within a couple of years this levels off. Some companies have also discovered unanticipated hidden costs, such as lawsuits for injuries related to use of exercise equipment.

On the other hand, many employees obviously do benefit from the programs their companies offer. If nothing else, it's convenient, for those who are so inclined, to exercise or attend a Weight Watchers class during their lunch break. In addition to workouts and workshops, the more complete programs offer true health-care services, such as cancer screenings and physical therapy for lower-back pain.

Whatever may eventually prove to be the effects on the physical condition of employees and the financial well-being of their employers, the deeper significance of corporate fitness initiatives lies in how they alter the relationship between employer and employee. When a company builds a fitness facility, it sends a message, even to staff who do not attend, about the intimate role the corporation claims in their lives. Notes Peter Conrad, a sociologist who recently completed a major study of corporate "wellness" programs: "Worksite health promotion programs with their focus on smoking, exercise, diet, blood pressure and the like are entering the domain of what has long been considered private life. Corporations are now increasingly concerned with what employees are doing off company time."

Some insurance companies have started to charge lower rates to people who exercise and keep their weight down, and before long employees' take-home pay may also be affected by whether they subscribe to the current physical ideals. Some leaders in the corporate health movement have recommended that those who take up exercise, lose weight, or stop smoking should be paid more than others. Even if this does not come to pass, an unfortunate effect of the corporate fitness movement is that employees who become sick or overweight are blamed for any ills that befall them.

Union leaders and advocates of occupational health legislation have expressed legitimate worry that all the emphasis on building muscles and dropping pounds is directing public attention away from a set of serious health problems corporate leaders are less eager to address. Two million Americans are disabled every year by work-related accidents and diseases, 33 million are injured by unsafe consumer products, and uncounted millions more suffer with allergies and illnesses caused by air and water pollution generated by industry.

The important point about these work-related hazards is that many can be eliminated; the technology is available to do so if corporate officers are willing to take the necessary steps. In contrast, our level of scientific knowledge is actually very limited on how to improve health by changing individuals' personal habits. Edmund Pellegrino, a medical ethicist, has noted that only in a limited number of areas do reasonably definitive studies exist. We can be fairly confident, for instance, that smoking increases the risk of cancer, and that wearing seat belts improves the odds of surviving an automobile accident. In other areas, the quality of the evidence, as well as the amount of disagreement among researchers, is much greater. These areas, Pellegrino indicates, include the degree of danger caused by cholesterol, stress, or lack of exercise.

Marshall Becker, head of the Department of Health Behavior and Health Education at the University of Michigan, recently chastised his colleagues in the health-promotion business for overzealousness. "Instead of waiting for larger and better studies or for some coalescence of scientific information," he writes, "we have rushed to recommend important modifications in individuals' life styles, with the following results: reasonably content people have had their fears aroused and feel compelled to attempt significant behavioral changes, attempts at which many (if not most) will fail; some advice is subsequently considered to have actually been *harmful;* and the public has become confused, and even skeptical, about public health advice, perhaps especially because we offer our contradictory advice sequentially."

The level of disagreement among authorities is especially great when it comes to exercise. How much we should get, and which types, remain unresolved questions.

ELEVEN

Exercise Won't Save You

Something incredible happened while I was writing this book: I became an exercise buff. Not because *I* changed, mind you. The culture did. Suddenly, an activity I had innocently pursued for years became the latest fitness craze.

I'm talking about walking.

According to surveys conducted by national sporting goods associations, 28 percent more Americans were walking for exercise in 1986 than the year before, while 12 percent fewer were running. At the same time, sales of running shoes dropped by $46 million. On Manhattan's trendy Upper West Side, a store called the Urban Hiker, Ltd., billing itself as "America's first walking store," opened its doors for business. Also in 1986, while *The Runner* merged with *Runner's World*, leaving only one national magazine for runners, *The Walking Magazine* premiered; and *Prevention*, which published a special issue and regular features on walking (and many pages of ads for walking shoes), started a national walking club. By year's end, the circulation of *The Walking Magazine* was half a million, and *Prevention*'s walking club boasted 56,000 members.

Then, in 1987, the national press got on the bandwagon.

That spring, you couldn't stroll past a magazine counter without bumping into headlines about walking—"Walkers Outpacing Joggers," read the front page of *USA Today;* "Why Walking's Hot, Running's Not," rhymed *Self;* "Get Walking, Get Healthy, A Total Guide," hyped *Vogue.* Editors competed with one another to give the flashiest advice on what to wear while walking, how to lose weight by walking, what to do when it rains, and whether or not to carry weights.

I felt like M. Jourdain, the character in Molière's *The Would-be Gentleman*, who exclaims, "Good heavens! For more than forty years I have been speaking prose without knowing it."

Why Americans Walk

When I asked some analysts in the sporting goods industry what had brought about this walking rage, they summed up the answer in one word: *demographics.* Baby boomers were hitting forty and needed something less grueling than running or heart-pounding aerobics; and concurrently, the number of elderly who were capable of and interested in exercise had grown quite large.

As reasonable as that may sound, neither of these facts really explains why it was that *walking* caught on. After all, just a few years ago, people responded to encroaching middle age by becoming *more* active, not less. The running craze was created by men in their forties and fifties, and high-impact aerobics were popularized by Jane Fonda, a woman in her forties. As for the elderly, many other exercise options might have captured their interest—dancing, biking, or swimming, to name a few.

As much as I'd like to attribute the walking boom to sociological causes, I can't. The major reason millions of Americans took up walking in the late eighties, I'm convinced, is that a public-relations wizard named Carol Cone engineered a brilliant campaign on behalf of one of her clients.

Why am I so sure that Cone is responsible? Because when I delved into scores of newspaper and magazine articles, advertisements, and pamphlets on walking, hoping to identify distinctive cultural themes in the walking movement, I found instead—a shoe company: Rockport, one of Cone's clients.

An event Cone dubbed "The 50/50 Walk for the Health of It," organized to promote Rockport's walking shoes, started it all. In 1984 Cone arranged for a man named Rob Sweetgall to spend fifty weeks walking across fifty states. Sweetgall had committed himself to fitness three years earlier, after several members of his family died of cardiovascular disease. By the time Cone and Rockport transformed him into a walker, he'd already developed something of a record for himself by running a 3,500-mile race and delivering public lectures about the rewards of exercise.

Still, when Cone first announced Sweetgall's walk, the press was indifferent. Reporters from the various news services and magazines dutifully showed up at a press conference Cone threw at Tavern on the Green, a restaurant in New York City's Central Park, in late 1984, but little came out in print.

Cone persisted. Once the walk was under way, Cone called her contacts in the media every week to inform them of Sweetgall's progress. To add a medical angle to the story, she flew Sweetgall back to Boston eight times during his trek and had him tested by James Rippe, a cardiologist at the University of Massachusetts, an institution that was another of Cone's clients at the time.

Not until Sweetgall had traipsed several thousand miles, and Rippe had documented improvements in his physical condition, did the tide finally turn. *USA Today* ran a page one story on Sweetgall, which prompted other newspapers, as well as radio and television stations, to take an interest. By the time it was all over, the 11,600-mile journey—which ended at New York's South Street Seaport with bands, balloons, and a cheering crowd that Cone had had bussed in—resulted in 500 interviews, with, among others, *Today*, *Good Morning America*, and *The Wall Street Journal*.

As soon as the hoopla died down, Cone, having created a spectacle in support of walking, needed next to create some legitimacy for this "new" form of exercise. To that end, she and Rippe came up with "The Rockport Walking Test," which measures a person's level of fitness solely on the basis of speed and heart rate during a one-mile walk. At a press conference in June 1986 at a YMCA in New York, Cone introduced the test, and within a matter of months Rockport had distributed 750,000 pieces of information about it. Variations have since appeared in national magazines.

Cone also established The Rockport Walking Institute, for which Sweetgall serves as education director and Rippe as research director. The institute gives out small grants for research on walking and maintains a scientific advisory board consisting of—in Cone's words—"thirty of the nation's leading exercise physiologists."

The institute has become a useful promotional tool. When one of its members or grantees comes up with a research finding on the healthfulness of walking, Cone makes sure it is publicized. She and Rippe also initiate articles in influential professional publications. For example, Rippe moderated a round-table discussion by several Ph.D.s associated with the institute and published an edited version in *The Physician and Sports Medicine*. One of the participants, a Wayne State University professor, is quoted as saying: "Some of the best advice you can give a friend is to 'take a walk.' Research clearly shows that walking is a viable cardiovascular conditioner. It is highly effective in improving aerobic capacity and reducing fat." Also in that article, a biomechanics expert testifies that there is a greater risk of musculoskeletal damage in running than in walking, and an exercise physiologist cites evidence that walking slows the progression of osteoporotic disease.

Rippe and company have placed articles in several other respected medical journals as well, thereby building what Cone refers to in her marketing memos as "an unimpeachable platform using medical/scientific research and medical experts."

The mass media have also taken the bait. *U.S. News and World Report* ran a cover story in 1986 on "new rules for exercise"; the magazine hyped up walking, spotlighted Rippe, and cited Rockport as the leading walking shoe. To get that kind of attention, Cone and Rippe talked with the magazine's writers and editors frequently and plied them with facts and quotes from members of the advisory board of the institute.

One result of all this publicity is that Rockport's sales increased from $18 million in 1983 to $120 million in 1986, when the company was sold to Reebok. (Meanwhile, Cone's own revenues increased from $80,000 in 1980 to $3 million in 1987.) Since that time, several other manufacturers have made major inroads into the walking-shoe market as well, and the market itself has continued to grow. According to industry analysts, Americans will spend $360 million on walking shoes in 1988, up by more than $100 million over 1987. By comparison, the projected market for running shoes in 1988 is $280 million.

Companies outside the sporting goods industry are even starting to cash in. Sanka, for instance, has sponsored a National Walking Week and provided a research grant to Rippe for a study of the psychological benefits of walking. As Cone says: "Walking is so wholesome, it's a wonderful thing to be associated with. The image that running conjures up is sweat and pain: marathoning. You can't talk when you run, it's individualistic. Walking conjures up fun, pleasure, family."

Cone is plainly proud of the Rockport campaign. Not only did it bring her commendations from the PR community and new accounts for her firm, but she believes she's provided a public service to boot. Until quite recently, says Cone, "a forty- or fifty-five-year-old woman who didn't look phenomenal would see these bright and bouncy young women on TV doing aerobics, who *do* look phenomenal, and she would feel there was no exercise for her to do. She wouldn't put on tight body clothes and jump up and down like that, because she's got bad knees or ankles or high blood pressure or she doesn't have good breasts.

"One of the things we did that I feel professionally very proud of is that we enfranchised millions of people to exercise by making it chic to walk."

But from the point of view of a veteran walker like me, this new veneration for walking is a mixed blessing. On the one hand, it's refreshing to be able to stride around town without incurring jeers from teenagers and superior glances from joggers. On the other hand, all the ballyhoo brings its own burdens. Time was, we walkers could just slip on a comfortable pair of shoes and go for a stroll whenever the spirit moved us. Now that we've been bombarded with magazine articles and shoe company ads, we don't dare. First we have to spend a day reading test reports on the 250 or so models of walking shoes currently on the market (in January 1986 there were twenty). Heaven help those of us who still amble along in running shoes; according to some of the ads for walking shoes, we actually risk injury.

Then there's the matter of supplementary exercises. Every walking program calls for warm-ups and cool-downs, and some prescribe so many flexibility, stretching, and strength exercises that the walk itself becomes almost an afterthought. Sweetgall and Rippe's book, *Fitness Walking*, actually advises special sit-ups and push-ups.

In the face of medical evidence that walking can be good for one's health only if performed properly, a *schedule* is also required. The walking books and articles contain tables that dictate distances, optimal number of walks per week, and heart rates. Prescriptions vary, depending on which research lab devised the table, but all require you to be at 12 to 20 miles per week, 4 miles per hour, and a heart rate of 70 to 80 percent of maximum within a dozen or so weeks from the time you start walking.

If all that obligation, time, and complexity don't scare away a potential or experienced walker, the hyperbole will. Walking has been touted as the new panacea. It "gives you a mental lift, a sense of well-being, accomplishment, and pride," according to *Fitness Walking*, and keeps you "active, well, and in control of your life."

"Walking is more than an exercise," Sweetgall and Rippe gush. "It is a ticket to appreciating the world around you."

Such overstatements are hauntingly reminiscent of those made some ten years earlier about running. An article in *Ladies' Home Journal* in 1978 avowed that running "helps you fall asleep in a wink, sleep like a baby," and that it "inspires creative flashes, fantasies, [and] clear thinking." George Sheehan said he was "born again" when he started to run at age forty-five. He called running "a physiologically perfect exercise." But most Americans who tried running did not find the promised spiritual or physical fulfillment; and the hundreds of thousands who injured themselves also did not find it so "physiologically perfect." Much the same will no doubt be true of walking: only a small percentage of converts will derive the sublime contentment promised in the books. Many will find walking nothing but boring and time-consuming.

Take Two Workouts and Call Me in the Morning

The fact is that some people find that their physical and mental states improve when they exercise, while others don't. A majority (estimates run from 50 to 90 percent) of those who join exercise programs drop out. Sports psychologists point to factors like inconvenience, the repetitiveness of most forms of exercise, and injuries. What may be more important, though, is the inevitable discovery that even if you are devoted and work hard, neither happiness nor health nor beauty is guaranteed.

The benefits of exercise have been grossly exaggerated at several periods in American history, and always on the basis of "medical evidence." Doctors in the 1830s promoted exercise as a cure for tuberculosis and as a preventive-health measure for sedentary wealthy women; others in the 1850s devised "movement cures" for everything from gout and constipation to "nervous afflictions"; a doctor in the 1870s recommended exercise to young men as an alternative to the nasty habit of masturbation; and a physician toward the end of that century proclaimed that exercise was a "microbe killer" which would protect people from contagious diseases.

Today, Dr. Kenneth Cooper, "the father of aerobics," carries on the tradition. Cooper has intimated in his books that exercise can alleviate constipation, depression, stress, migraines, varicose veins, coronary artery disease, poor eyesight, blemishes on the skin, and diminished intellectual performance. Cooper and other apostles of exercise never tire of telling stories about people so sluggish and depressed they seem dead, who lose twenty pounds, look ten years younger, make more money, and are no longer anxious or depressed after only a few months of aerobic conditioning.

But read a large number of articles and books about exercise, as I have done, and you discover that only one thing is certain: despite a tremendous amount of research and optimism, we have very little knowledge about exercise that is uncontroversial.

Experts can't even agree on *how many* people exercise. A Gallup survey in the March 1987 issue of *American Health* reported a huge upturn between 1984 and 1986 in the number of Americans who exercise regularly—from 54 to 69 percent. That same month, a story in *The New York Times* reviewed other surveys that showed little or no change in the number of exercisers. A few months later, an issue of the *University of California Wellness Letter* pointed to government health surveys from 1985 showing that only 40 percent of Americans exercise regularly.

In a single folder of press materials sent to me by Carol Cone, I found the number of fitness walkers in America estimated variously at 42 million, 55 million, and 70 million.

Nor is there accord about *what sort* of exercise we ought to engage in. Partisans get hot under the collar when they argue the merits of one form of exercise over another. For example, Fred Stutman, a physician and aficionado of walking, warns that jogging produces heart attacks and destroys virtually every major organ of the body. He's not wild about other popular options, either. "Calisthenics," he writes in his book *Walk Don't Die*, "cripple by breaking bones, pulling muscles, stretching ligaments, and ripping tendons. Aerobic dancing damages feet, ankles, knees, hips, and spines by pneumatic-drill type bouncing. . . . As far as I'm concerned, aerobics are awful and calisthenics are crazy."

Kenneth Cooper, on the other hand, advocates a wide range of exercise options including running, walking, and aerobic dance. On one fundamental point, though, the two doctors concur. Both say that to protect your health, you need to engage in *aerobic* exercise—exercise that increases your heart rate for a sustained period of time so that your body burns extra oxygen.

But some authorities dispute even the necessity for that. In an article in *The Physician and Sports Medicine,* a group of epidemiologists from the University of Pittsburgh complain that "for too long the U.S. public has been told that if an activity program was not designed for aerobic conditioning, it would not reap health benefits." In actuality, they contend, "the epidemiological evidence suggests risk of disease is likely to decrease by merely increasing activity irrespective of its aerobic nature."

Not a single study has demonstrated that vigorous exercise reduces the risk of heart attack, according to these researchers. Rather, what the studies show is that people who are more *active* have lower rates of heart disease—whether the activity comes from participation in aerobic exercise or otherwise.

The epidemiologists also refute one of Kenneth Cooper's favorite claims: that by decreasing your pulse rate through aerobic exercise, you protect yourself against heart attacks. In one of his books, Cooper says he persuaded his wife to start an exercise program by spelling out to her the significance of his fifty-beat-per-minute resting heart rate as compared to her eighty beats. "While we're asleep tonight," Cooper quotes himself as having said, "your heart is going to beat about 10,000 times more than mine will. Even though our hearts are pumping the same *amount* of blood, it takes your heart that much more work and effort to do the job because you're not in condition. You're just going to wear out faster than I will.''

Against arguments of this sort, the Pittsburgh epidemiologists cite several studies showing that a person's resting heart rate is not a good predictor of whether he or she will have a heart attack. Cooper's heart-rate conceit sounds plausible, but there's no compelling evidence behind it.

The informed disagreement doesn't end here. Try to determine *how much* exercise (of whatever sort) you should perform each week, and you're sure to experience mental fatigue. Obviously, if your goal is to get in shape for ski season or a mountain-climbing expedition, or to keep up with your three-year-old child or grandchild, you're going to have to do more than is necessary for health reasons alone. But if those sorts of things don't interest you—and for uncounted millions of Americans, they don't—if your concern is protection against heart disease, how much exercise do you need?

Widely published recommendations from sports physiologists and cardiologists range from three times a week for fifteen minutes each at 60 percent of your maximum heart rate, to five times for forty minutes at 80 percent pump rate.

Some doctors call for considerably *more* exercise. Arthur Leon, a cardiologist at the University of Minnesota, has said, ''Our research suggests that one hour's worth of even moderate exercise is what really reduces the heart attack risk.'' Last I heard, George Sheehan was still promoting running thirty miles a week as optimal for improving one's health without great risk of injury. (Kenneth Cooper, on the other hand, has lately recommended twelve to fifteen miles of running. He previously advocated fifteen to twenty miles, but after some bone and heel problems of his own he said: "I've changed my mind. I'm running less and performing better.")

At the other end of the track—in a lounge chair, you might say—are authorities who claim that very little exercise is plenty. According to *Executive Fitness Newsletter*, for example, the busy executive can stay fit even exercising only on weekends. Time-Life Books, meanwhile, sells a volume promising "a complete aerobic workout you can perform while sitting at your desk."

Then there's Cornell University cardiologist Henry Solomon, who claims that the equivalent of a leisurely one-mile walk per day is sufficient. "You must be truly sedentary," Solomon writes, "a slug who sits or lies about all day, or barely crawls from bed to breakfast, to car and desk and back again, to be at any risk from inactivity. No one who has to push a vacuum cleaner, play ball with the children or keep the lawn mowed is that inactive."

Marathon runners might demur, but the voices of moderation have evidence to back them up. A study in the *Journal of the American Medical Association* found that the blood chemistry of men with heart disease improves significantly if they walk just a little more than a mile and a half a few times a week. Doctors writing in the *American Journal of Epidemiology* reported that any regular physical activity results in a lower rate of heart attacks and a better survival rate after heart attack. In fact, the group of University of Pittsburgh researchers quoted above have asserted: "The epidemiological research might suggest that gardening three times a week is as effective as running for preventing heart attack."

Such conclusions contrast sharply with some that were widely publicized in the seventies and early eighties in favor of more vigorous exercise. For example, *The New England Journal of Medicine* published a report in 1975 stating that marathon running provided immunity from heart disease. The physicians who reached this conclusion wrote that they "were unable to document a single death from ischemic heart disease among marathon finishers of any age." Unsurprisingly, once skeptics started looking into the matter, definitive autopsy reports emerged showing atherosclerosis in samples of marathon runners.

Henry Solomon suggests that the lower rates of heart disease and better cholesterol levels in championship runners are easily attributable to factors other than exercise. In general, these athletes have lower cholesterol levels before they start to run, and come from families with healthier hearts than do nonrunners.

Such self-selection—choosing to pursue particular forms of exercise or lines of work because you're in good enough shape to do so—is a fundamental problem with much of the research literature on exercise. Lower rates of heart disease have been documented in physically active workers such as railway and dockyard employees, for instance, but they may have been less at risk before ever picking up a shovel. They appear to have been thinner to begin with, with less heart disease in their families, and relatively low cholesterol and blood-pressure levels. Being healthy in the first place, they may have selected careers that made use of their physical prowess.

Selection problems also confound the findings from perhaps the most famous study showing a connection among exercise, health, and longevity. Generally referred to as "the Paffenbarger study," after the senior researcher, Ralph Paffenbarger, this research made the front page of *The New York Times* in 1986 and is routinely cited as definitive evidence that moderate exercise increases longevity. Paffenbarger, an exercise devotee who has run several 100-mile races, based his conclusion on questionnaires mailed back by male alumni of Harvard University, ages thirty-five to seventy-four. In every age group, fewer of the more active men suffered heart attacks or died during the sixteen-year study.

Yet the study is inconclusive because it could not rule out that the more active subjects were different in important physiological and sociological ways from those who were generally sedentary. Quite possibly, active people are blessed with the sorts of genes, marriages, or careers that dispose them to live longer, with or without exercise, and that may also provide them with greater energy and interest in exercise.

Healthier people may also *describe* themselves as more physically active than do less healthy people. If so, Paffenbarger's results could be misleading, since he estimated the amount of physical activity his subjects performed by asking them how many blocks they walked, how many stairs they climbed, and how much time they spent playing sports each week.

How many of us really know the accurate answers to those questions? If those who were healthier paid more attention to stairs, miles, or tennis games than those who were less healthy, then the findings of the study are based on fictitious numbers. Paffenbarger himself has admitted that his question about the number of stairs climbed was readily misinterpreted because some people gave the number of *flights* of stairs they'd climbed, while others reported every single step.

To my eye, the more interesting findings from Paffenbarger's research are ones that received far less publicity than the correlation between exercise and longevity. Paffenbarger also learned that death rates for his Harvard alums as the result of various illnesses, including heart disease and cancer, were about half those of the general U.S. white male population. Given that the median income for Harvard grads is roughly twice the national average, and that about one-fifth are millionaires, a moral of the Paffenbarger study would seem to be: to protect yourself from disease, go to a fancy university and make lots of money.

On the other hand, suicide rates for Paffenbarger's sample were 50 percent *higher* than national averages. Perhaps graduates of Harvard are more in need of suicide prevention programs than exercise programs.

Catchy Correlations

What we often forget, when we read in the newspaper about a study like Paffenbarger's, a study heralding a relationship between exercise and good health, is the difference between correlation and causation. The fact that one variable frequently follows another does not prove that the first brought about the second. Hospitalization and death are highly correlated, for instance, but rarely does a stay in the hospital cause someone to die. More commonly, the cause of a person's death is a third factor, the same one that caused the hospitalization: a serious illness.

Then, too, a great many phenomena are correlated simply by accident. Their rates of occurrence may rise or fall together, but they have no real relationship. Statistics professors love to cite such examples as the strong correlation once found between how much Presbyterian ministers in Massachusetts are paid and the price of a bottle of rum in Havana. No rational person would seriously argue that lowering the salaries of these preachers would bring down Cuban booze costs. But in the area of health promotion, every time a correlation is discovered between a particular diet or exercise plan and heart disease or cancer, health promoters are quick to recommend that people change their behaviors.

Marshall Becker, the health education specialist from the University of Michigan, illustrates the folly of this approach.

240

He cites a study that found a correlation between low cholesterol levels and high death rates from accidents, suicide, and homicide. "Does adhering to a low-cholesterol diet," asks Becker facetiously, "make one wish to commit murder? To *be* murdered? While I have not been able to construct a theory to account for this new threat, I now lunch only with people who eat steak and french fries."

We Americans like to believe that we're masters of our own fate, that if we heed the advice of scientists as reported in the newspapers, we'll live to a ripe old age and save a fortune on doctor bills. In reality, says Becker, "The domain of personal health over which the individual has direct control is *very* small when compared to heredity, culture, environment, and chance. Nor is health promotion the panacea for rising health-care costs."

Will exercise save your life? The most honest, informed answer seems to be that a truly sedentary life-style is dangerous, but beyond that the picture is unclear.

Many health benefits from exercise have been advertised, but none is guaranteed. Even a statement as seemingly self-evident as *exercise leads to weight loss* requires qualification. A hearty workout does burn calories, and many people have successfully lost weight that way, but let's not forget the thousands of underweight people who take up exercise every year in order to *gain* weight. And although contemporary diet plans consistently recommend exercise as an appetite suppressor, some dieters find—as did William Banting, author of a bestselling nineteenth-century diet book—that a workout actually makes them hungrier.

Even the Fit Get the Blues

Exercise has been particularly oversold in the area of mental health. We've all read magazine articles recommending a regular shot of aerobic activity as an antidote for whatever ails our psyches. Some therapists actually prescribe exercise as a form of treatment.

Look to the research literature, however, and you find stacks of contradictory findings. In some experiments, participants' levels of depression and anxiety decrease when they exercise, in others they remain essentially unchanged, and in some they actually increase. And a recent Gallup poll suggests that those who exercise are no happier overall than those who do not. Equal numbers of exercisers and nonexercisers surveyed agreed with the statement: "In general, I am very satisfied with my life the way it is right now."

A few years ago John Hughes, an epidemiologist at the University of Minnesota, reviewed more than 1,100 studies on the effects of exercise on mood, personality, and thinking. Only twelve of the studies, he learned, met minimal standards of scientific acceptability—meaning that they involved controlled experiments with ten or more subjects who were randomly assigned to exercise programs lasting at least three weeks.

Of the various findings reported in those twelve credible studies, only one-third indicate a positive psychological benefit from exercise. Another third are negative, and the final third show mixed results. Hughes concludes that "the enthusiastic support of exercise to improve mental health has a limited empirical basis and lacks a well-tested rationale."

His assertion is supported by another review of the research literature, this one appearing in *The American Psychologist*, which states, "In general, studies of physical fitness effects on psychological health are poorly designed."

Many of the studies consist of quick, single-shot experiments that discover that people's mental states improve after they exercise. This is not very strong evidence in favor of the value of exercise, though, because we know from other research that almost *any* reasonable change in people's behaviors will make them feel better momentarily. Psychologists have also found they can improve people's emotions by having them sit up straight or smile. A researcher at Skidmore College produced elevated moods and feelings of greater energy in her subjects just by changing their gait during a three-minute walk around a room.

The more important question is whether exercise helps in the longer run. Unfortunately, many of the studies that suggest it does are comparisons of athletes and nonathletes. They do suggest that athletes are better able to deal with stress; but who's to say whether this ability is the result of exercise rather than some other factor? Athletes and nonathletes differ in a number of ways that give athletes an advantage in times of trouble. Personality research shows that athletes are more adventurous, self-controlling, strong-willed, and extroverted. And athletic adults tend to be college-educated and have higher-than-average incomes, according to sociological studies.

A better research design starts with nonathletes and observes how they change over some weeks or months in an exercise program. Yet in those studies too, it's difficult to verify that exercise is the critical factor that improves a person's mental health. Social influences may be more important. Quite possibly, what cheers people up and allays their anxieties is the opportunity to meet new people, or simply to get out of the house.

A further concern when interpreting positive findings from such experiments is the likelihood that those who volunteer for an exercise program are already motivated to improve. Even if they're not, they may experience a "placebo effect"— they feel better after exercising because they believed they would.

One category of studies which largely avoids these methodological problems looks at the effects of exercise on people with Type-A personalities. By definition, the subjects in these experiments are busy and skeptical types, and hence unlikely to want more social contacts or to hold positive expectations about the experiment at hand.

As a matter of fact, some early studies that placed Type-A people in exercise programs did seem to show that regular workouts helped to cool down their anger and aggressiveness. But a recent experiment with 107 managers from two large Canadian corporations indicates otherwise. The men, Type A's all, were put in three groups. For ten weeks, one group was given aerobic exercise; the second was instructed in stress-management techniques such as communication skills, problem solving, and muscular relaxation; and members of the third (the control group) were put through what they thought was a beneficial weight-training program, but which was actually at a level too limited to provide significant benefits.

The results were dramatic. Type-A behaviors of men given stress-management instruction decreased substantially. They grew markedly less hurried, hostile, loud, competitive, and explosive. Even the control group improved somewhat. But those in the exercise group showed virtually no change. (Nor did exercise improve their physiological readings such as blood pressure and heart rate, by the way.)

Whether an exercise regimen contributes to your well-being will depend largely on how you view exercise and how it fits into your life. If you enjoy a good run or hike or game of tennis, then engaging in those activities can help you burn off some tension and forget your troubles. But if you're a confirmed couch potato who doesn't find exercise pleasurable, you're not likely to reap many psychological rewards by working up a sweat.

Perhaps the Type-A executives in the Canadian study gained more from relaxation sessions than from aerobic work-outs because they consider exercise *itself* a stressful and competitive activity.

Christine, the forty-year-old executive, made this point in describing her views about stress. "I don't accept the premise that you can fight fire with fire," she told me. "You fight fire with water. I work in a hard business environment, and I don't want extracurricular activities that are also work. I want to soften how hard I have to work. What calms me down is two hours in front of the television set.

"Later in life, after I've put away enough money that I can keep up my standard of living by occasional consulting, my life will be about softer things. When I'm sixty I'd like to be living in the country spending my days gardening, watching old movies on the VCR, and making pottery. Then I might take up exercise. It would become my work. I might take on jogging or something just to provide a counterpoint for all that soft stuff."

Because Christine is self-confident and attractive, she's able to *just say no* to exercise. Most of us are more susceptible to social pressures. The reasons we're susceptible are not always as obvious as they may seem.

TWELVE

The Secret Agenda

In 1985, at the height of the fitness craze, when record numbers of Americans were exercising and eating low-fat diets, *Psychology Today* conducted a large national survey on how people feel about their bodies. The magazine had run a similar study back in 1972, before most Americans started working out and eating less meat.

A comparison of the results reveals an alarming fact: we are doing more for our bodies but liking them less. On every dimension studied, Americans grew *less* satisfied with their bodies between the early 1970s and mid-1980s. More are displeased with how much they weigh, and more dislike each of the regions of their anatomy the survey inquired about—face, chest, abdomen, and lower torso. The exercise revolution hasn't even made us feel stronger. While 30 percent of women in 1972 were dissatisfied with their muscle tone, the number jumped to 45 percent in 1985. For men it rose from 25 to 32 percent.

My interviews around the country reinforce the conclusion that all our efforts to beautify and condition our bodies have not made us, as a nation, any happier with the way we look. Why, then, do we spend tens of billions of dollars every year to remake ourselves?

I suggest it's because the work we do on our bodies affords us something far more valuable than satisfaction when we look in the mirror. Whether or not it actually improves our appearance, body work gives us something we desperately crave in this post-Watergate, post-Vietnam, post–Ollie North age. It allows us to feel morally pure.

Just listen to the language we use. If we eat fattening food or skip our daily workout we tell our friends we've "been bad." Certain chocolate desserts we refer to as "wicked" or "sinful." And we believe fat people to be "weak-willed," and say they should be "ashamed of themselves."

Admittedly, going to a health club or ordering broiled fish at a restaurant does not *sound* like a great moral deed. Customarily, when we think of moral behaviors, what comes to mind are the sorts of traits the Bible talks about—honesty, charity, the avoidance of jealousy.

Nonetheless, proper eating and exercise are perfect candidates for moral acts in a modern society. Nearly a hundred years ago the social theorist Emile Durkheim saw that judgments of right and wrong were coming to be based on the findings of scientists rather than on the teachings of religious leaders or philosophers. He also noted that moral acts were increasingly viewed by people not as dictates from on high but as choices made freely and based on their informed belief that such acts were in their own best interest.

Over the course of the twentieth century, a true remoralizing of our culture has taken place in the name of science and personal choice. Ironically, in the process we've reverted to some of the least appealing beliefs found in so-called primitive societies. For example, many Americans now accept the notion that a person becomes ill not because of natural physical processes but because of immoral action.

That belief was held throughout most of the world, including Europe, until the acceptance of modern medical science.

Following on Pasteur's studies of bacteria, health and disease came to be understood in terms of the control of germs. But now that bacterial diseases are no longer a major threat in Europe and America, and we're confronted with chronic and largely incurable maladies, we have reintroduced ritual and blame into the picture. If someone is stricken with a heart attack or cancer, many of us do not wonder about the complex environmental and genetic factors that led to the disease. Instead we suspect the illness was the person's own fault: he or she should have exercised or eaten properly.

For some illnesses, patients are blamed outright. A book on diabetes, released in 1982 by a respectable publishing house, claims that 90 percent of adult-onset diabetics are overweight and that "these overweight diabetics can control their diabetes condition and sometimes *eliminate* it by using diet and exercise to maintain a proper weight," since three out of four diabetics could have prevented their illness.

Numerous low-calorie diets have been administered to diabetics over the years, on the assumption that diabetes is, in the words of a leading physician of the 1920s, "largely a penalty of obesity, and the greater the obesity, the more likely is Nature to enforce it." Actually, though, the epidemiology of diabetes is more complex than that. The condition may sometimes be caused by obesity, but in other cases it is the other way around: diabetes contributes to obesity. Various factors aside from fatness are known to produce the symptoms of the disease.

What's more, the fact that a person's diabetes is controllable by means of diet and exercise does not logically imply (and no scientist has shown) that had the person weighed less he or she would not have developed the condition. Yet the author of that 1982 book moralizes that one reason for increases in the number of diabetics "is that many people are careless and lazy about their health."

More Than a Glance Is Cast

I've come across some shocking statistics in my research, but here is the one I find most alarming of all. A Gallup poll taken in 1986 found that 93 percent of Americans agree with the statement: "If I take the right actions, I can stay healthy." The corollary to this belief is, of course, that if you do become ill, you must not have taken care of yourself.

These convictions, although factually inaccurate, are widely believed because they have the power of moral rightness and medical authority behind them.

I'm not suggesting that people shouldn't try to maintain a healthy life-style, or that they shouldn't encourage others to do so. But that's not all that's going on here. A great deal of guilt-mongering and victim-blaming is taking place— and class and racial biases have come to the fore.

When we condemn fatness or sedentary ways, who are we actually condemning? Which segments of the American population exercise the least, eat the most sugar and fat, and have the highest obesity rates? The lower social classes and ethnic minority groups. By some estimates, as many as half of lower-class women are obese, compared to less than 10 percent of upper-class women. Likewise, the wealthier a person is, the more exercise that person is likely to take; about one-third of people with low incomes get little or no formal exercise, compared to less than one-fifth of those with high incomes. And higher percentages of blacks than whites are overweight and sedentary.

The health moralists of the late twentieth century are furthering the cause of prejudice and discrimination against these groups as surely as the temperance crusaders who stigmatized immigrants and laborers early in the century on account of their drinking.

Comparisons of that sort greatly offend "wellness" advocates, who see themselves as liberals and progressives. Yet they defend their position with the standard paternalistic cant: *"We're just telling them for their own good.* Once people are given the facts about exercise and diet, they'll thank us for saving their lives."

Baloney. A lack of information in poor and working people's lives is not what keeps them from healthier living. Time and money are.

Neither fancy equipment nor a health club membership is required in order to exercise—although both certainly help in the motivation department—but anyone who exercises needs time. Some low-income people do have time on their hands and use it to exercise—unemployed men in the ghetto and inmates, for instance. But the apartment-bound mother of three or the person who works swing shifts or holds two jobs to make ends meet is not likely to have much chance for a workout. They're lucky if they get adequate sleep and three meals a day.

As for a healthy diet—that costs both money and time. Poorer people are fatter and more prone to certain diseases largely because, in the United States, the foods that are cheapest and fastest to buy and prepare also tend to be salty, greasy, starchy, or sugar-filled.

Working people often *are* aware that their diet and level of fitness might shorten their lives. They do watch TV, after all, and visit doctors' offices. A recent study suggests that black and poor patients are lectured to by doctors and other health professionals about their eating habits more often than are other Americans.

The least privileged in our society, then, are not generally ignorant about these matters. Rather, many of them reject the arguments they've heard, and others accept them but give higher priority to other concerns in their lives.

When Robert Crawford, a political scientist, asked a cross-section of Chicagoans to define the word *health*, he discovered two contradictory ways in which people think about this concept:

On the one hand, middle-class professionals typically defined health in terms of discipline and self-control—you're healthy if you deny yourself cigarettes and red meat and maintain an exercise routine. For these people, reports Crawford, "Health is not a given, nor is it simply a result of good luck or heredity, two frequently mentioned alternatives. Neither is it believed to be an outcome of normal life activities, such as one's work, upbringing or current lifestyle. Health must be achieved."

This is the Horatio Alger view of health; one becomes healthy the same way one becomes wealthy: through hard work and wise investments of time and money. Crawford notes that some people in his study who subscribed to this philosophy of health did not actually engage in the "healthful" activities they cited because, they said, they were too busy doing other things, like raising children or making ends meet.

On the other hand, many of the working-class and poor people Crawford interviewed rejected the very notion of self-control as the route to good health. Perhaps because they and their families had not found much validity to the American myth that hard work and self-restraint bring prosperity, they didn't accept its extension into the realm of health either.

Their contrary view was succinctly expressed by one interviewee, who defined *health* this way: "It's being able to do what you want to do when you want to do it. It's nice to be able to do things. It's nice to be able to eat and drink what you want, not worrying about being overweight or sticking to the diet that the doctor told you to stick to."

From this perspective, health comes not from restraining

251

yourself and doing things you don't really want to do, but the opposite, from letting go and enjoying life. There's enough stress, obligation, and control imposed on us in our work lives, these people point out, that we don't need still more during our leisure hours. People who hold this view are often aware that they might live longer if they changed some of their habits. But they wonder whether the reward outweighs the costs. As one of the men I interviewed asked, "Which is better? To raise some hell when you're young enough to enjoy it, or to hang around an extra couple of years in a nursing home when you're ninety years old?"

The gulf is wide between the affluent and the less-than-affluent on these matters—and once during my travels I found myself smack in the middle of it. I was at a roadside restaurant near Atlanta. At the table to my left, three young couples dressed in K-Mart's finest were having a grand time, the men talking to the men and the women to the women. In the fifteen minutes between the time the waitress took their orders for cheeseburgers, fried chicken, fries, and onion rings and delivered it all, the six of them went through sixteen Michelobs.

Each was a good thirty to fifty pounds heavier than insurance tables would recommend, and, just as their food arrived, in walked two other couples who were their mirror opposites. About the same age (late twenties or early thirties), the new-comers, who sat themselves at the table to my right, were lean, fit, and decked out in Adidas and Esprit cotton exercise wear.

The menu didn't please this second party very much; they complained among themselves that almost everything was greasy. When the waitress came to take their orders, each asked for a salad and glass of water.

Throughout my own meal—an attempted compromise of

tuna sandwich, onion rings, and one bottle of beer—I over-
heard hushed comments about each table from the other. A
woman to my right registered the belief that menus should
carry warning boxes like those on cigarette cartons: "This
food causes obesity, heart disease, and cancer." One of the
men to my left asked his buddies if they'd heard the joke
about how you know when you've run over a jogger in your
pickup truck—something about credit cards flying onto the
windshield.

The healthy foursome departed as soon as the salads were
finished. The beer drinkers were still there after I left, working
on slices of banana cream pie. As for me, I drove off imagining
that it would be easier to wangle a Soviet-American détente
than to mediate between these two tables of young Americans
on the subject of how to care for their bodies.

Several of the blue-collar people I interviewed expressed
the opinion that something must be bothering middle-class
folks who fixate on their bodies. In their own experiences
with friends and family, when people pay too much attention
to their bodies, usually something else is wrong. People fuss
over their hair or their aches and pains when they're having
troubles with their husband, their wife, or their boss.

To some extent this explanation is defensive because it
permits a person to dismiss as unimportant a status symbol
he or she does not own. In our society, the body has become
an important instrument for registering a superior position.
A nonfat, nonsmoking Nautilus body differentiates its owner
from the unwashed, uneducated masses. That's one reason
why certain ghetto dwellers emulate or exaggerate it.

Nonetheless, there is truth to the claim that hidden motivations lie behind the urge to work on one's body. And when
whole segments of a society preoccupy themselves with their

bodies, clearly something besides their good health is at stake.

I have suggested that that something is morality. It is more, though, than a *personal* morality. The deeper motivation behind the fitness rage is a hunger for *public* morality . . . a hunger that, ultimately, no amount of exercise or healthy eating can satisfy.

Can Your Body Save America?

Liberal middle-class Americans may not agree with many of the government directives over the last eight years. But they're unlikely to find fault with the official objective, released in 1980 by the Public Health Service, that 60 percent of all adults under age sixty-five take up regular vigorous physical exercise by 1990; and that by that same year, the number of overweight Americans be reduced to 10 percent of men and 17 percent of women.

There's a long way to go to reach these figures, given Public Health Service estimates that, as of 1985, less than 8 percent of Americans got enough exercise and 24 percent of both men and women were significantly overweight.

Still, with national objectives of this sort and corporations telling their employees that their good health is a key to the future of American industry, the ideal images in the magazines are much more than just pictures of pretty bodies. They're symbols of hope for the future of America.

To appreciate the prodigious meaning the lean and healthy body holds for much of the middle class today, it helps to recall the roots of the contemporary fitness craze. The movement's founding father was none other than John F. Kennedy, who held that for a nation to be strong, its populace must be fit. "The physical vigor of our citizens," he proclaimed, "is one of America's most precious resources. If we allow it to dwindle and grow soft then we will destroy much of our ability to meet the great and vital challenges which confront our people."

In speaking that way (and he did so frequently), Kennedy revived "a two-millennia-old tradition of seeing the individual body as a sign—both as metaphor and as source—of the health or infirmity of the larger social body," as literary critic Catherine Gallagher has put it. By means of this reasoning, Kennedy sold the nation on an idea which is, on the face of it, preposterous in the age of automation—that the nation is fueled by the strength and energy of its citizens.

Kennedy initiated major new exercise programs in the schools and the military, thus creating a generation of young Americans brought up to believe in the value of fitness, and to associate it, subconsciously perhaps, with the last president they really admired.

On adults of his time, however, Kennedy's fitness crusade had little immediate and practical impact; they did not take up exercise in record numbers until after Kennedy's death. In fact, his assassination, along with those of his brother and of Martin Luther King, Jr., were themselves important stimuli of the fitness revolution, which really took off with the jogging craze in the second half of the 1960s.

The early joggers were middle-aged, middle-class men who used running as spiritual therapy. Though they were doing well financially, these men had grown discouraged. Their national heroes had been murdered, their careers and marriages seemed spiritually unrewarding, and their fathers and friends were dying of heart disease in record numbers.

Muriel Gillick, a social historian, has suggested that two factors led to the running movement. "First," she says, "was the realization that modern medicine, for all its sophistication, could not prevent death. Even the coronary care unit, one of the great technological developments of the '60s, saved at most a few lives; and 60 percent of deaths from heart attacks occur before the victims ever reach medical attention."

The second factor, Gillick argues, was political. "The collapse of the liberal consensus—the belief that the strength and virtue of America had created peace abroad and harmony at home—coming on top of a shattered faith that American medicine could render the world safe from disease, led to the view that America was morally sick, in need of spiritual renewal."

There were several diverse responses to the death of the liberal consensus. Young people grew their hair long, women stopped wearing bras, blacks went searching for their history. And middle-aged white men took their passions into the streets by jogging. "The pursuit of physical fitness was seen by some as a means by which individuals could improve America," notes Gillick. "By ridding us of the stress and tension, the competitiveness and sleeplessness which are ruining our society, so the argument goes, running can help us pull ourselves up by our bootstraps."

In due course, women also took up fitness as a vehicle of political and moral reformation. "The new female consciousness that has developed over the last decade," declared Jane Fonda in her first workout book, "extends to our right to physical as well as economic, political and social equality."

That book, lately a mainstay at suburban garage sales, mixes instructions on how to perform chest-to-knee stretches and extended scissor kicks with stirring essays on the perils of food additives and air pollution. The connections Fonda made were not always the most logical, but her books led to a great deal of talk among feminists about the empowerment exercise and proper diet can bring, and to articles in *Ms.* magazine extolling bodybuilders and other physically strong women as the new role models.

□

Recently, though, there are signs the tide may be turning. Advocates of walking and other exercises of the late-1980s still glorify the health-and-beauty benefits of working up a good sweat, but they don't promise political restitution. And some critics have gone so far as to suggest that remaking one's body may be a repressive rather than liberating act.

In *Beauty Bound*, feminist psychologist Rita Freedman describes the women's fitness revolution as "simply another insidious beauty oppression." It's not a revolution at all, says Freedman, but a throwback. "The rash of new fitness magazines for women preach a kind of enlightened narcissism based on health. But their pages are filled with fashion tips, weight-loss testimonials, an emphasis on vanity, and a demand for self-sacrifice. On the whole, the message is not so very different from the one found in traditional fashion magazines. It is a familiar appeal for masochistic make-over in the name of beauty, this time cloaked in a mesomorphic cover-up."

Then there's the view of Ivan Illich, who in 1975 had declared modern medicine a threat to public health. Upon reading his book *Medical Nemesis*, as well as other critiques of the medical establishment published in the early 1970s, many people concluded that the health-care system is so dangerous that they'd better do everything possible to keep themselves well. But lately Illich has revised his diagnosis. After observing the fads of the past decade, Illich recently wrote: "Today's major pathogen is, I suspect, the pursuit of a healthy body." He criticized "those who conceive of themselves as 'producers' of their bodies."

The word is gradually seeping out that we can't rescue ourselves, either individually or as a nation, by way of our bodies. By getting physical we have kept our moral feelings alive through the difficult post-Vietnam and Reagan eras, but now we must find other, more realistic options.

People are starting to recognize that the ideal body isn't so ideal after all, that it stands not only for beauty and health but also for false hopes and prejudices. That knowledge may be disheartening at first, but it also frees us—to exercise and eat in ways that match our own needs rather than the dictates of the latest fad, to cope better with illness and aging, and to turn our moral energies in more meaningful directions.

Notes

ONE: DAILY LIVES

Vogue statistics: *Harper's,* March 1986, p. 11.

Estimate of dollars spent on body enhancement: Alison Thresher, "Girth of a Nation," *Nation's Business,* December 1986, pp. 50–51; also industry reports.

Limitations of surveys: Serge Lang, *The File,* New York, Springer-Verlag, 1981; Aaron Cicourel, *Method and Measurement in Sociology,* New York, Free Press, 1964; and Howard Schwartz and Jerry Jacobs, *Qualitative Sociology,* New York, Free Press, 1979.

A note to my academic colleagues regarding methodology: For this book I have opted to work within the tradition of "first-person sociology." For a discussion of this approach, see Bob Blauner, "Problems of editing 'first-person' sociology," *Qualitative Sociology,* volume 10, 1987, pp. 46–64. For an important argument on the validity of individual case studies as a sociological research tool, see J. Clyde Mitchell, "Case and situational analysis," *The Sociological Review,* volume 31, 1983, pp. 187–211. In some of my earlier works, where my goals have been different—for instance, where particular lives are not the focus of the study, and where generalizations have been made directly from a set of interviews—I have taken a more positivistic approach to sampling and to the construction of interview schedules. See, for instance, my and Julia Loughlin's *Drugs in Adolescent Worlds,* New York, St. Martin's, 1987, and my and Bruce Berg's "How Jews avoid alcohol problems," *American Sociological Review,* volume 45, 1980, pp. 647–64. In those works, the interviewees' remarks are reproduced verbatim, unlike here, where I have followed the lead of Robert Coles and Studs Terkel and have edited the interview transcripts.

On the paucity of body vocabulary: Elaine Scarry, *The Body in Pain*, New York, Oxford University Press, 1985, pp. 12–19. Scarry notes the very small vocabulary available for describing pain.

TWO: THE POWER OF THE IMAGE

The *Ms.* issue: April 1987.

The feminist revolt against beauty was never complete: Judith Levine, "Our bodies, our clothes: Fashion and feminism," *Utne Reader*, August/September 1986, pp. 38–46.

On looking and destroying: Otto Fenichel, "The scoptophilic instinct and identification," in *The Collected Papers of Otto Fenichel*, New York, Norton, 1953, pp. 373–97; quotes appear on pp. 377 and 390–91. For hints at an alternative interpretation, see Scarry, *The Body in Pain*, cited earlier, p. 354, n. 176.

Early history of images of the body: Liam Hudson, *Bodies of Knowledge*, London, Weidenfeld and Nicolson, 1982, chapter 4.

Shift from hearing and touching to seeing: Donald M. Lowe, *The History of Bourgeois Perception*, Chicago, University of Chicago Press, 1982, chapters 1 and 5; Hillel Schwartz, *Never Satisfied*, New York, Free Press, 1986, pp. 162–64; quote appears on p. 163.

Photos and photo collections in Victorian America: Robert Bogdan, *Freakshow: Sociological Encounter with History*, Chicago, University of Chicago Press, 1988, chapter 1.

Anne Hollander quote: *Seeing Through Clothes*, New York, Viking, 1978, p. 154.

The body in a product-oriented society: Bryan S. Turner, *The Body and Society*, London, Blackwell, 1984; Richard Sennett, *The Fall of Public Man*, New York, Vintage, 1978; Peter Freund, *The Civilized Body*, Philadelphia, Temple University Press, 1982; and the book this chapter is titled after, Annette Kuhn, *The Power of the Image*, London, Routledge & Kegan Paul, 1985. Because this is a book for a general audience, my discussion of what social theorists refer to as "commodification" is necessarily truncated and devoid of technical terms. In a paper prepared for the 1988 Society for the Study of Symbolic Interaction Stone Symposium I offer a more theoretical and sustained argument.

Regarding early department stores: Lois Banner, *American Beauty*, New York, Knopf, 1983, pp. 32–36.

Mike Featherstone quote: "The body in consumer culture," *Theory, Culture and Society*, volume 8, 1982, pp. 18–33; quote appears on p. 28. On status and image see also Erving Goffman, *Gender Advertisements*, Cambridge, Harvard University Press, 1979; and *Behavior in Public Places*, New York, Free Press, 1963.

Christopher Lasch quote: *The Minimal Self*, New York, Norton, 1984, pp. 29–30.

Product and image study: Stanley Baran and Vincent Blasko, "Social perceptions and the by-products of advertising," *Journal of Communication*, volume 34, 1984, pp. 12–20.

On the image-based economy: Jean Baudrillard, *For a Critique of the Political Economy of the Sign*, St. Louis, Telos Press, 1981; and *The Mirror of Production*, St. Louis, Telos Press, 1975; Stuart Ewen, "The political elements of style," in Jeffery Bucholtz and Daniel B. Monk (eds.), *Beyond Style*, New York, Précis, 1984, pp. 125–33.

Study of twentieth-century advertisements: William Leiss, Stephen Kline, and Sut Jhally, *Social Communication in Advertising*, New York, Methuen, 1986; quote appears on p. 231. On the personalization of products in advertisements in the early years, see also Roland Marchand, *Advertising the American Dream*, Berkeley, University of California Press, 1985, chapter 5. Note also that of the many ads reproduced in that book, the vast majority that include people feature highly attractive subjects.

Ajax adman: quoted in *Syracuse Style*, July 23, 1986, p. 1. Further, sociologist Varda Leymore has shown how advertisements convey powerful messages equating "the use and the non-use of the product with one of the great dilemmas of the human condition." In the ads Leymore studied, mothers were called upon "to use baby foods since they proclaim Life, detergents since they are signified by Good, frozen vegetables for their reincarnation of Nature, dehydrated exotic meals for the secret of Knowledge" (Varda Leymore, "Structure and persuasion: The case of advertising," *Sociology*, volume 10, 1982, pp. 337–89).

Per-second cost of television ads: *U.S. News and World Report*, October 20, 1986.

Memory of faces: Joseph J. Fleishman, Mary Lou Buckley, et al., "Judged attractiveness in recognition memory of women's faces," *Perceptual and Motor Skills*, volume 43, 1976, pp. 709–10; Michael J. Baker and Gilbert A. Churchill, "The impact of physically attractive models on advertising evaluations," *Journal of Marketing Research*, volume 14, 1977, pp. 538–55.

Why beautiful models sell: Murray Webster and James Driskell, "Beauty as Status," *American Journal of Sociology*, volume 89, 1983, pp. 140–65; Gordon Patzer, "Source credibility as a function of communicator physical attractiveness," *Journal of Business Research*, volume 11, 1983, pp. 229–41; Ellen Bercheid and Elaine Walster, "Physical attractiveness," in L. Berkowitz (ed.), *Advances in Experimental Social Psychology*, New York, Academic Press, 1974, volume 7, pp. 157–215; J. Bassili, "The attractiveness stereotype," *Basic and Applied Social Psychology*, volume 2, 1981, pp. 251–52. Another hypothesis for why beautiful bodies are effective in ads is based on evidence that most people view themselves as more attractive than they really are. (See Gordon L. Patzer, *The Physical Attractiveness Phenomena*, New York, Plenum, 1985, pp. 34–37.) That argument implies that we identify with great-looking people because we see ourselves as one of them. Perhaps. But surveys reveal that most of us are also dissatisfied and insecure about how we look.

Financial advantages of attractiveness: J. Richard Udry and Bruce Eckland, "Benefits of being attractive: Differential payoffs for men and women," *Psychological Reports*, volume 54, 1984, pp. 47–56; R. P. Quinn (1978), cited in Elaine Hatfield and Susan Sprecher, *Mirror, Mirror: The Importance of Looks in Everyday Life*, Albany, State University of New York Press, 1986.

Experiment on choice of men to meet: Nicholas Gallucci, "Effects of men's physical attractiveness on interpersonal attraction," *Psychological Reports*, volume 55, 1984, pp. 935–38.

On stereotypes and realities about the attractive: Hatfield and Sprecher, *Mirror, Mirror,* cited earlier, chapters 2, 8, and 12; Nicholas Gallucci and Robert Meyer, "People can be too perfect," *Psychological Reports*, volume 55, 1984, pp. 351–60; Gallucci, "Effects of men's physical attractiveness on interpersonal attraction," cited earlier; Marshall Dermer and Derrel L. Thiel, "When beauty may fail," *Journal of Personality and Social Psychology*, volume 31, 1975, pp. 1168–76.

Advantages and disadvantages of beauty for women: Rita Freedman, *Beauty Bound,* Lexington, Massachusetts: Lexington Books, 1986; Udry and Eckland, "Benefits of being attractive," cited earlier.

Effects of attractiveness while growing up: *Parents*, October 1983, pp. 72–74; Katherine Hildebrandt and Teresa Cannan, "The distribution of caregiver attention in a group program for young children," *Child Study Journal*, volume 15, 1985, pp. 43–55; *Jet*, April 29, 1985, p. 27.

Self-monitoring men: Mark Snyder, Ellen Bercheid, and Peter Glick, "Focusing on the exterior and the interior: Two investigations of the initiation of personal relationships," *Journal of Personality and Social Psychology*, volume 48, 1985, pp. 1427–39.

Studies of models in ads: Alice Gagard, "From feast to famine: Depiction of ideal body type in magazine advertising, 1950–1984," paper presented at the annual meetings of the American Academy of Advertising, March 1986; Paula England, Alice Kuhn, and Teresa Gardner, "The ages of men and women in magazine advertising," *Journalism Quarterly*, volume 58, 1981, pp. 468–71.

Monika Schnarre quote: *People*, June 30, 1986.

Cover photo credit: *Mademoiselle*, December 1986.

Susan Sontag quote: *On Photography*, New York, Farrar, Straus & Giroux, 1973, p. 24.

THREE: NO BODY IS PERFECT

Paulina Porizkova quote: *Gentleman's Quarterly*, September 1986, p. 367.

Images and anorexia: Linda L. Smith, "Media images and ideal body shapes: A perspective on women with emphasis on anorexia," paper presented at the conference of the American Educators in Mass Communication and Journalism, Memphis, 1985.

Experiment on beauty ads and young women: Alexis S. Tan, "TV beauty ads and role expectations of adolescent female viewers," *Journalism Quarterly*, volume 56, 1979, pp. 282–88.

Effects of images on how men see women: Douglas T. Kenrick and Sara E. Gutierres, "Contrast effects and judgments of physical attractiveness: When beauty becomes a social problem," *Journal of Personality and Social Psychology*, volume 38, 1980, pp. 131–40; James B. Weaver, Jonathan Masland, and Dolf Zillmann, "Effect of erotica on young men's aesthetic perception of their female sexual partners," *Perceptual and Motor Skills*, volume 58, 1984, pp. 929–30.

Studies of grade-school children: Jeffrey Zaslow, "Fourth-grade girls these days ponder weighty matters," *The Wall Street Journal*, February 11, 1986, pp. 1, 20; *The Chronicle of Higher Education*, November 19, 1986, pp. 11–12.

Body image of the disfigured: Seymour Fisher, *Development and Structure of the Body Image*, New Jersey, Erlbaum, 1986, p. 265. See also Marcia Millman, *Such a Pretty Face*, New York, Norton, 1977.

Overweight as more significant for women: D. M. Garner et al., "Cultural expectations of thinness in women," *Psychological Reports*, volume 47, 1980, pp. 483–91; Orland W. Wooley et al., "Obesity and Women: II. A neglected feminist topic," *Women's Studies International Quarterly*, volume 2, 1979, pp. 81–92; Thomas Horvath, "Correlates of physical beauty in men and women," *Social Behavior and Personality*, volume 7, 1979, pp. 145–51.

Dissatisfaction with weight: Hatfield and Sprecher, *Mirror, Mirror*, cited earlier, pp. 204–5; Alexander Frazier and Lorenzo K. Lisonbee, "Adolescent concerns with physique," *School Review*, volume 58, 1950, pp. 397–405.

Discrimination against obese women: Helen Canning and Jean Mayer, "Obesity: Its possible effect on college acceptance," *The New England Journal of Medicine*, volume 275, 1966, pp. 1172–74. In a survey of 1,660 Manhattan residents, the long-term effects of discrimination came out clearly. Obese women—far more often than either thinner women or obese men—had fallen below the social class level of their parents (Phillip B. Goldblatt, Mary E. Moore, and Albert J. Stunkard, "Social factors in obesity," *Journal of the American Medical Association*, volume 192, 1965, pp. 1039–44). See also Hatfield and Sprecher, *Mirror, Mirror*, cited earlier, pp. 209–13.

Obese prejudice against the obese: Judith Rodin and Joyce Slochower, "Fat chance for a favor," *Journal of Personality and Social Psychology*, volume 29, 1974, pp. 557–65.

Medical students study: David B. Herzog et al., "Eating disorders and social maladjustment in female medical students," *Journal of Nervous and Mental Disease*, volume 173, 1985, pp. 734–37.

FOUR: MOTHERS AND DAUGHTERS

Importance of social background for body shape and physical comportment: Sanford M. Dornbusch, "Norms for thinness among adolescent females," and B. Diane Hayes et al., "Exercise and well-being," both papers presented at the annual meetings of the American Sociological Association, New York, 1986; Marilyn F. Steele and John H. Spurgeon, "Body size, body form, and nutritional intake . . . ," *Growth*, volume 47, 1983, pp. 207–16; S. Leonard Syme and Lisa F. Berkman, "Social class, susceptibility, and sickness," *American Journal of Epidemiology*, volume 104, 1976, pp. 1–8; Edwin and Aloo Driver, "Social class and height and weight in metropolitan Madras," *Social Biology*, volume 30, 1983, pp. 189–204; Tadeusz Bielicki et al., "The influence of three socio-economic factors on body height in Polish military conscripts," *Human Biology*, volume 53, 1981, pp. 543–55; Catherine E. Ross and John Mirowsky, "Social epidemiology of overweight," *Journal of Health and Social Behavior*, volume 24, 1983, pp. 288–98; Turner, *The Body and Society*, cited earlier.

Heritability of obesity: Stanley Garn, Marquisa LaVelle, and Jeffrey Pilkington, "Obesity and living together," *Marriage and Family Review*, volume 7, 1984, pp. 33–47; Judith Gordon and Alice Tobias, "Fat, female and the life course," *Marriage and Family Review*, volume 7, 1984, pp. 65–91; Jill Haraway, "Structural assessment of families with obese adolescent girls," *Journal of Marital and Family Therapy*, volume 12, 1986, pp. 199–201.

Anorexia, the family, and social change: Susan Bordo, "Anorexia nervosa: Psychopathology as the crystallization of culture," *The Philosophical Forum*, volume 17, 1986, pp. 73–104; Marlene Lodahl-Boskind, "Cinderella's stepsisters: A feminist perspective on anorexia nervosa and bulimia," *Signs*, volume 2, 1976, pp. 342–56; Neville Golden and Ira M. Sacker, "An overview of the etiology, diagnosis, and mistreatment of anorexia nervosa," *Clinical Pediatrics*, volume 24, 1984, pp. 209–14; Hilde Bruch, *Eating Disorders*, New York, Basic Books, 1973; and *The Golden Cage: The Enigma of Anorexia Nervosa*, New York, Vintage, 1979.

Man's Man: Leonard L. Glass, "Man's Man/Ladies' Man: Motifs of Hypermasculinity," *Psychiatry*, volume 47, 1984, pp. 260–77.

Nancy Chodorow quotes: *The Reproduction of Mothering*, Berkeley, University of California Press, 1978, pp. 166, 195. See also Signe Hammer, *Daughters and Mothers: Mothers and Daughters*, New York, Signet, 1976, p. 14.

Lucy Rose Fischer's study: *Linked Lives: Adult Daughters and Their Mothers*, New York, Harper & Row, 1986.

Lena Wright Myers's study: "Social support systems for black women," in *Black Women: Do they Cope Better?* Englewood Cliffs, New Jersey, Prentice-Hall, 1980, pp. 26–40.

Effects from overcontrolling or underinvolved mothers: Fisher, *Body Image*, cited earlier, pp. 374, 644–45.

Parents and teachers' expectations about attractive daughters: David Landy and Harold Sigall, "Beauty is talent: Task evaluation as a function of the performer's physical attractiveness," *Journal of Personality and Social Psychology*, volume 29, 1974, pp. 299–304; Cookie W. Stephan and Judith Langlois, "Baby beautiful: Adult attributions of infant competence as a function of infant attractiveness," *Child Development*, volume 55, 1984, pp. 576–85; Gerald R. Adams and Paul Crane, "An assessment of parents' and teachers' expectations of preschool children's social preference for attractive or unattractive children and adults," *Child Development*, volume 51, 1980, pp. 224–31.

Freud quote: cited in Fischer, *Linked Lives*, cited earlier, p. 2.

FIVE: FATHER, FOOD, FITNESS

Alexis de Tocqueville's observations: *Democracy in America*, New York, Doubleday, 1969, pp. 592–94, 600–603; quote appears on p. 601.

Ideal womanhood and hysteria: Carroll Smith-Rosenberg, "The hysterical woman: Sex roles and role conflict in 19th century America," *Social Research*, volume 39, 1972, pp. 552–78.

Anorexics and exercise obsessives: Alayne Yates, Kevin Leehey, and Catherine Shisslak, "Running: An analogue of anorexia?" *The New England Journal of Medicine*, volume 308, 1983, pp. 251–55; Bordo, "Anorexia nervosa" cited earlier; Turner, *The Body and Society*, cited earlier, chapter 8; M. R. Combs, "By food possessed," *Women's Sports*, 1982, pp. 11–17.

Mothers, and anorexia and bulimia: Steven W. Emmett (ed.), *Theory and Treatment of Anorexia Nervosa and Bulimia*, New York, Brunner-Mazel, 1985, see especially chapters 5, 7, and 11; Kim Chernin, *The Hungry Self*, New York, Harper & Row, 1985, quote appears on p. 56.

Susan and O. Wayne Wooley's observations: "Thinness mania," *American Health*, October 1986, pp. 68–86.

Susie Orbach quotes: "Visibility/invisibility: Social considerations in anorexia nervosa—a feminist perspective," in Emmett (ed.), *Theory and Treatment*, cited earlier, pp. 127–38.

The jogger study: reported in a New York Times News Service story of February 7, 1986. Other literature on exercise addiction: Stanton Peele, *How Much Is Too Much*, Englewood Cliffs, New Jersey, Prentice-Hall, 1981; William P. Morgan, "Negative addiction in runners," *The Physician and Sports Medicine*, volume 7, 1979, pp. 57–70.

Judith B. Elman's story: "The loneliest of the long-distance runners," *Runner's World*, volume 21, July 1986, pp. 34–39.

Blair Sabol quote: *The Body of America*, New York, Arbor House, 1986, p. 49.

Women's magazines study: Marjorie Ferguson, *Forever Feminine*, London, Heinemann, 1983, pp. 68–69.

On horror films: *Screen*, a special issue on "Body Horror," volume 27, January–February 1986.

Self-control and hysteria: Smith-Rosenberg, "The hysterical woman," cited earlier.

Signe Hammer's description: *Passionate Attachments: Fathers and Daughters in America Today*, New York, Rawson, 1982.

Substitute sons: Fischer, *Linked Lives,* cited earlier, p. 35.

Susan and O. Wayne Wooley quote: "Thinness mania," cited earlier, p. 73.

Daughters of alcoholics: Elizabeth Stark, "Forgotten victims: Children of alcoholics," *Psychology Today,* volume 21, 1987, pp. 58–62.

Mary Douglas's hypothesis: "The two bodies," *Natural Symbols,* New York, Pantheon, 1970, pp. 65–81. This explanation derives from the work of Emile Durkheim. See especially his *Elementary Forms of Religious Life,* New York, Macmillan, 1915; and *Suicide,* Illinois, Free Press, 1951.

Women in the labor force: Statistics are from Bureau of the Census and Department of Labor reports for 1985 and 1986; Jo Freeman, *Women: A Feminist Perspective,* Palo Alto, California, Mayfield, 1984; and Robert K. Richards, "The declining status of women . . . revisited," *Sociological Focus,* 1986, pp. 315–32.

SIX: MEN AND MUSCLES

Muscles in American history: Rupert Wilkinson, *American Tough,* New York, Harper & Row, 1986; quotes from the salesman's manual appear on pp. 38–39.

Bernarr Macfadden quote: Harvey Green, *Fit for America,* New York, Pantheon, 1986, chapter 9.

Studies of adolescent boys: Larry Tucker, "Relationship between perceived somatotype and body cathexis of college males," *Psychological Reports,* volume 50, 1982, pp. 983–89; Larry A. Tucker, "Effect of a weight-training program on the self-concepts of college males," *Perceptual and Motor Skills,* volume 54, 1982, pp. 1055–61; Luella W. Cole and Irma N. Hall, *Psychology of Adolescence,* New York, Holt, Rinehart & Winston, 1970; Joan Hemmer and Douglas Leiber, "Tomboys and sissies: Androgynous children," *Sex Roles,* volume 7, 1981, pp. 1205–12; Carol L. Martin, "Why are tomboys and sissies evaluated differently?" paper presented at the meetings of the American Psychological Association, 1985. See also Shere Hite, *Report on Male Sexuality,* New York, Knopf, 1981, pp. 10–16.

Muscles as signs of masculinity: Nancy Huston, "The matrix of war: Mothers and heroes," in Susan R. Suleiman (ed.), *The Female Body in Western Culture,* Cambridge, Massachusetts, Harvard University Press, 1986, pp. 119–36. See also Desmond Morris, *Bodywatching,* New York, Crown, 1985, p. 137; E. Anthony Rotundo, "Body and soul: Changing ideals of American middle-class manhood, 1770–1920," *Journal of Social History,* volume 16, 1983, pp. 23–38.

Mark Goodson's article: "Lousy at Sports," *The New York Times Magazine,* May 11, 1986, p. 48.

Muscles and self-esteem: Ellen Berscheid, Elaine Walster, and George

Bohrnstedt, "Body image," *Psychology Today*, 1973, p. 121; Larry A. Tucker, "Muscular strength and mental health," *Journal of Personality and Social Psychology*, volume 45, 1983, pp. 1355–60.

Charles Gaines quote: "Hulk triumphant," *Esquire*, June 1986, p. 118.

Study of personals advertisements: A research assistant and I analyzed the contents of one issue each of the following newspapers and magazines: *Intro* and *Matchmaker Matchmaker* of Los Angeles; *New York; The Village Voice; The New York Review of Books; The Queens Chronicle;* a Jewish newspaper from upstate New York; *The National Review;* a local newspaper in Fort Lauderdale, Florida; and *City Limits* of London. We selected these to obtain a cross section of audiences and regions, but as a glance at the list suggests, the sample is tilted in favor of the two coasts. We analyzed 272 ads placed by males, and 215 placed by females. All of the ads analyzed for this study were by apparent heterosexuals who sought heterosexual mates. Two-thirds of the advertisers were thirty to forty-five years old.

Motivations for exercise: Hillel Ruskin and Boas Shamir, "Making bodies beautiful: Male and female participation in a dance exercise aerobics program," paper presented at the annual meetings of the American Popular Culture Association, 1986; Barbara Johnson, "Motivation as a factor affecting males' participation in physical activity during leisure time," *Society and Leisure*, volume 7, 1984, pp. 141–61; Larry Tucker, "Effect of weight training on self-concept," *Research Quarterly for Exercise and Sport*, volume 54, 1983, pp. 389–97; Alan M. Klein, "Pumping irony: Crisis and contradiction in bodybuilding," *Sociology of Sport Journal*, volume 3, 1986, pp. 112–33.

Continuity of exercise over the lifespan: Barry McPherson, "Sport participation across the life cycle," *Sociology of Sport Journal*, volume 1, 1984, pp. 213–30; Eldon Snyder and Elmer Spreitzer, *Social Aspects of Sport*, Englewood Cliffs, New Jersey, Prentice-Hall, 1978; and "Lifelong involvement in sport as a leisure pursuit," *Quest*, volume 31, 1979, pp. 57–70; G. Lawrence Rarick, *Physical Activity: Human Growth and Development*, New York, Academic Press, 1973.

High school athletics and later status: Frank M. Howell, Andrew W. Miracle, and C. Roger Rees, "Do high school athletics pay?" *Sociology of Sport Journal*, volume 1, 1984, pp. 15–25.

SEVEN: POWER AND VANITY

A note about the argument in this chapter: Some authors have claimed, unconvincingly I think, that attractiveness is much less important to males than females. One who gives empirical evidence is Rita Freedman in *Beauty Bound* (cited earlier, see pp. 1, 2, 10, 29, 130–43, 202). She cites a survey in which women admitted they were interested in how men look but felt it socially unacceptable to admit as much. She also notes studies showing that women underestimate their attractiveness, while men overestimate theirs.

These studies might just as well be used, however, as arguments *against* Freedman's view that attractiveness is more important to women. If women gawk secretly, men may not have to suffer catcalls on the street, but their appearance is taken into account nonetheless. In a sense, men are put at a disadvantage in this exchange, since they get less feedback about how they look. Maybe that is one reason why men overrate their looks.

Harry Benson quote: *People*, March 30, 1987, p. 4.

National survey on satisfaction with appearance: Thomas F. Cash, Barbara A. Winstead, and Louis H. Janda, "The great American shape-up," *Psychology Today*, volume 20, April 1986, pp. 30–44.

My survey at Syracuse University: 154 undergraduate students were selected at random in 1986 and were administered a questionnaire. The comparison numbers for the women in the sample: 54 percent said they were "very concerned" about their appearance, and 2 percent said they were "not concerned."

New Mexico State University study: Larry A. Tucker, "Relationship between perceived somatotype and body cathexis of college males," *Psychological Reports*, volume 50, 1982, pp. 983–89.

Survey of senior citizens in Philadelphia: Laura N. Gitlin, "Body image as an enduring concern: Older adults' perceptions of their appearance," paper presented at a meeting of the Society for the Study of Social Problems, Washington, 1985.

Personals in the mid-1970s: Albert A. Harrison and Laila Saeed, "Let's make a deal: An analysis of revelations and stipulations in lonely hearts advertisements," *Journal of Personality and Social Psychology*, volume 35, 1977, pp. 257–64; and C. Cameron, S. Oskamp, and W. Sparks, "Courtship American style: Newspaper ads," *The Family Coordinator*, volume 26, 1977, pp. 27–30.

Personals in the mid-1980s: This refers to the analysis I conducted, which is described in the notes for the previous chapter. Other findings from my study indicated that the more things change, the more they stay the same. While 58 percent of the men who placed ads in the mid-eighties requested an attractive partner, only 16 percent of the women did. Apparently, it's still considered inappropriate for women to size up men overtly on the basis of their appearance. The relative importance of men's fiscal status has also remained constant. Just 3 percent of the men listed affluence as something they sought in a mate, while 26 percent of the women mentioned it as a trait they were after.

Male attractiveness and self-esteem: Fisher, *Body Image*, cited earlier, p. 129; Gerald R. Adams, "Physical attractiveness, personality, and social reactions to peer pressure," *Journal of Psychology*, volume 96, 1977, pp. 287–96.

Value of attractiveness for males and their mates: Jeffrey S. Nevid, "Sex differences in factors of romantic attraction," *Sex Roles*, volume 11, 1984, pp. 401–11; B. Gillen, "Physical attractiveness: A determinant of two types of goodness," *Personality and Social Psychology Bulletin*, volume 7, 1981, pp.

277–81; Kim Strane and Carol Watts, "Females judged by attractiveness of partner," *Perceptual and Motor Skills*, volume 45, 1977, pp. 225–26; Arie Nadler, Rina Shapira, and Shulamit Ben-Itzhak, "Good looks may help: Effects of helper's physical attractiveness and sex of helper on males' and females' help-seeking behavior," *Journal of Personality and Social Psychology*, volume 42, 1982, pp. 91–99.

Hiring study: M. Heilman and L. Saruwatari, "When beauty is beastly: The effects of appearance and sex on evaluations of job applicants for managerial and nonmanagerial jobs," *Organizational Behavior and Human Performance*, volume 23, 1979, pp. 360–72.

Women are viewed as looking better: Roger L. Terry and Elizabeth Macklin, "Accuracy of identifying married couples on the basis of similarity of attractiveness," *Journal of Psychology*, volume 97, 1977, pp. 15–20; John Cross and Jane Cross, "Age, sex, race and the perception of facial beauty," *Developmental Psychology*, volume 5, 1971, pp. 433–439.

Cynthia S. Rand and Judith A. Hall's remarks: "Sex differences in the accuracy of self-perceived attractiveness," *Social Psychology Quarterly*, volume 46, 1983, pp. 359–63.

Mary Douglas's observations: *Purity and Danger*, New York, Praeger, 1966.

EIGHT: COUPLES

Why a husband would want a fat wife: Richard B. Stuart and Barbara Jacobson, *Weight, Sex and Marriage*, New York, Norton, 1987, chapter 4.

Robert D. Hess and Gerald Handel's analysis: "The family as a psychosocial organization," in Gerald Handel (ed.), *The Psychosocial Interior of the Family*, New York, Aldine, 1985, pp. 33–46; quotes appear on pp. 36–38.

Chris Pepper Shipman's book: *I'll Meet You at the Finish!* New York, Human Kinetics Publications, 1987.

NINE: IN THE NAME OF HEALTH

Harvard Medical School cosmetic surgery study: Myron L. Belfer, John B. Mulliken, and Thomas C. Cochran, "Cosmetic surgery as an antecedent of life change," *American Journal of Psychiatry*, volume 136, 1979, pp. 199–201.

Minneapolis cosmetic surgery center ad: for Alt Cosmetic Surgery Center, in *Twin Cities*, February, 1987, p. 26.

Albert Lippert quote: "Weight Watchers: Think thin and grow fat," *Nation's Business*, September 1978, p. 100.

Happiness among the obese: Diane Hayes and Catherine E. Ross, "Body and mind: The effect of exercise, overweight and physical health on psychological well-being," *Journal of Health and Social Behavior*, volume 27, 1986, pp. 387–400; Cash et al., "The great American shape-up," cited earlier.

Skepticism about "thinner is better": Cheryl Ritenbaugh, "Obesity as a culture-bound syndrome," *Culture, Medicine and Psychiatry*, volume 6, 1982, pp. 347–61; Schwartz, *Never Satisfied*, cited earlier, chapter 11; Reubin Andres, "Effect of obesity on total mortality," *International Journal of Obesity*, volume 4, 1980, pp. 381–86; Jennifer B. Donovan, "Weighty debate: Is fat so bad?" *USA Today*, January 6, 1986, p. E1; Werner J. Cahnman, "The stigma of obesity," *The Sociological Quarterly*, volume 9, 1968, pp. 283–99; Jane Brody, "Research lifts blame from many of the obese," *The New York Times*, March 24, 1987, p. C1.

Statistics on weight-loss programs and dieters: information from H. J. Heinz Co. (owner of Weight Watchers); Rebecca Fannin, "Shape up!" *Marketing and Media Decisions*, February, 1986, pp. 54–60; N. R. Kleinfield, "The ever-fatter business of thinness," *The New York Times*, September 7, 1986, pp. C1, C28; Angela Kinamore, "Trimming down to size," *Essence*, May 1987, p. 142; "Beware 'yo-yo' dieting," *USA Today*, January 6, 1986, p. E1; Hilary Rosenberg, "Nutri/System after the fall," *Financial World*, August 15, 1983, pp. 38–39; Millman, *Such a Pretty Face*, cited earlier, p. xi; Cheryl Russell, "That fat feeling," *Utne Reader*, August/September 1986, p. 39; Cash et al., "The great American shape-up," cited earlier.

Plastic surgery statistics: information from the American Society of Plastic and Reconstructive Surgeons; Elizabeth Kaye, "The case for a polyester face," *Savvy*, January 1986, pp. 42–45; Cash et al., "The great American shape-up," cited earlier; Gary M. Kaplan, "Putting on a happier face," *Nation's Business*, volume 74, 1986, pp. 40–41.

California plastic surgery clinic ad: in Diana Dull and Candace West, "The price of perfection: A study of the relations between women and plastic surgeons," paper presented at the meetings of the American Sociological Association, 1987.

Robert M. Goldwyn quotes: "Plastic surgeons on the make," *Plastic and Reconstructive Surgery*, volume 75, 1985, pp. 251–52.

Potential risks and health-care drains of plastic surgery: Doug Lefton, "Ads entice patients to cosmetic surgery," *American Medical News*, volume 28, September 6, 1985, pp. 3, 17; Sheldon Rosenthal, "Malpractice: Through the looking glass," *Annals of Plastic and Reconstructive Surgery*, volume 9, 1982, pp. 326–29; Ann Scheiner, "My face-lift: A cautionary tale," *Ms.*, November 1986, pp. 58, 63, 81–82; Kay Williams, "The high costs of looking young," *Money*, April 1985, pp. 67–76; Elin Jones, "Cosmetic surgery," *Essence*, April 1986, pp. 36–41; Bowen Northrup, "Doctors doing cosmetic work scrap over turf," *The Wall Street Journal*, February 26, 1987, p. 25.

Patients prosper from cosmetic surgery: Fisher, *Body Image*, cited earlier, pp. 147–50; J. A. Hollyman, J. H. Lacey, et al., "Surgery for the psyche," *British Journal of Plastic Surgery*, volume 39, 1986, pp. 222–24; Ellen Berscheid and Steve Gangestad, "The social psychological implications of facial physical attractiveness," *Clinics in Plastic Surgery*, volume 9, 1982, pp. 289–95; J. Reich,

"The surgery of appearance," *Medical Journal of Australia*, volume 2, 1969, pp. 5–13.

Personality improvement study: S. Michael Kalick, "Toward an interdisciplinary psychology of appearances," *Psychiatry*, volume 41, 1978, p. 243–53.

TEN: HEALTH CLUB HAWKERS

Michel Foucault's history: *Discipline and Punish*, New York, Pantheon, 1977, pp. 135–36.

Health club industry statistics: information from the International Racquet Sports Association; Better Business Bureau publications; Bill Richards, "Misshapen identities," *The Wall Street Journal*, April 21, 1986, p. 150; Patricia Kilburg and Dev Strischek, "Lending to health clubs," *Journal of Commercial Bank Lending*, volume 67, 1984–85, pp. 8–20.

Health club problems and legislation: personal letters and publications from officials at Better Business Bureaus, the National Association of Attorneys General, and state attorneys general; Federal Trade Commission publications including "Report of the presiding officer on proposed trade regulations rule: Health spas 16 CFR part 443," April 1979; "FTC staff takes it easy on health-club operations," *Consumer Reports*, volume 50, 1985, p. 643; Sylvia Porter, "Exercise judgment when joining health spas," *Syracuse Herald Journal*, February 7, 1986.

Certification: "Instructor certification: Making fitness programs safer," *Parks and Recreation*, volume 21, December 1986, pp. 24–29.

Percent exercising and dropping out: Carl J. Caspersen, Gregory Christenson, and Robert Pollard, "Status of the 1990 physical fitness and exercise objectives," *Public Health Reports*, volume 101, number 6, 1986, pp. 587–92; Kilburg and Strischek, "Lending to health clubs," cited earlier.

Hospital programs: information from the American Hospital Association; articles and advertising in *Optimal Health*, September/October 1986 (Alan Schwartz quotes appear on p. 64); Galen Cole et al., "A systems perspective for hospital-based health promotion," *Optimal Health*, November/December 1986, pp. 24–28; "The health industry finally asks: What do women want," *Business Week*, August 25, 1986, p. 81.

Fort Smith, Arkansas, fitness center: mentioned in *Optimal Health*, May/June 1987.

Corporate fitness statistics and general information: Carlos Castillo-Salgado, "Assessing recent developments and opportunities in the promotion of health in the American workplace," *Social Science and Medicine*, volume 19, 1984, pp. 349–58; "Fit for the job," *The Economist*, October 12, 1985, p. 80; Christine Howe, "Establishing employee recreation programs," *Journal of Physical Education, Recreation, and Dance*, volume 54, 1983, pp. 52–53; Robert Crawford, "You are dangerous to your health," *Social Policy*, volume 8, 1978, pp. 11–20; Jack McCallum, "Everybody's doin' it: Getting into the fitness business, that is," *Sports Illustrated*, December 3, 1984, pp. 72–86. Quoted brochure is from Rodale Press.

On evolving American views of exercise: John Betts, "Mind and body in early American thought," *Journal of American History*, volume 54, 1968, pp. 787–805; Green, *Fit for America*, cited earlier; and Peter Levine, "The promise of sport in antebellum America," *Journal of American Culture*, volume 2, 1980, pp. 623–34.

Characteristics of program joiners: Peter Conrad, "Who comes to work-site wellness programs: A preliminary review," *Journal of Occupational Medicine*, volume 29, 1987, pp. 317–20.

1983 Tenneco study: William B. Baum, Edward Bernacki, and Shan Tsai, "A preliminary investigation: Effect of a corporate fitness program on absenteeism and health care cost," *Journal of Occupational Medicine*, volume 28, 1986, pp. 18–22.

For an informative study of a health-promotion program: Jennie Kronenfeld et al., "Health behavior and attitude change: Does health promotion really matter?" paper presented at a meeting of the Society for the Study of Social Problems, 1985.

Problems and prospects of worksite programs: Peter Conrad, "Wellness in the workplace: Potentials and pitfalls of worksite health promotion," *Millbank Quarterly*, volume 65, 1987 (in press); Steven N. Blair et al., "A public health intervention model for work-site health promotion," *Journal of the American Medical Association*, volume 255, 1986, pp. 921–26; Meg Fletcher, "Wellness plan savings are difficult to measure," *Business Insurance*, volume 21, February 16, 1987, pp. 26, 28; Roy J. Shephard, "Motivation: The key to fitness compliance," *The Physician and Sports Medicine*, volume 13, 1985, pp. 88–101.

Lawsuits: John H. Schultz, "Workers compensation claims: A deepening cloud over employee recreation services," *Journal of Physical Education, Recreation, and Dance*, volume 54, 1983, pp. 56–57.

Health-care services offered in corporate programs: Andrew J. Brennan, "Health and fitness boom moves into corporate America," *Occupational Health and Safety*, volume 54, 1985, pp. 38–45.

Points of critique regarding corporate health programs: Patricia Vertinsky, "Risk benefit analysis of health promotion," *Quest*, volume 37, 1985, pp. 71–83; Peter E. S. Freund, *The Civilized Body*, Philadelphia, Temple University Press, 1982, pp. 32–35; Peter Conrad, "Health and Fitness at Work," *Social Science and Medicine* (in press); and "Wellness in the workplace," cited earlier.

Insurance discounts: Glenn Kramon, "The wellness discount plans," *The New York Times*, September 22, 1987, p. D2; "Tying insurance rates to health risk," *Optimal Health*, September/October 1987, p. 14.

Pay-differentials suggestion: "Health risk appraisals," *Optimal Health*, November/December 1986, p. 29.

Work-related disabilities: *Statistical Abstracts of the United States:* 1987, Washington, D.C., U.S. Bureau of the Census.

Injuries from consumer products: "Product safety: It's no accident," Washington, D.C., National Inquiry Information Clearinghouse, 1986.

Edmund D. Pellegrino's remarks: "Health promotion as public policy," *Preventive Medicine,* volume 10, 1981, pp. 371–78.

Marshall H. Becker quote: "The tyranny of health promotion," *Public Health Reviews,* volume 14, 1986, pp. 15–25; quotes appear on pp. 15–16.

ELEVEN: EXERCISE WON'T SAVE YOU

Walking and running survey: Tom Topousis, "Walkers outpacing joggers," *USA Today,* May 28, 1987, p. 1.

Running-shoes statistics: Sporting Goods Manufacturers Association.

Walking magazine and club statistics: Margaret Carlin, "It takes more than shoes for walking," *Syracuse Herald American Stars Magazine,* January 18, 1987, p. 18.

Information about Carol Cone Communications and the Rockport campaign: telephone interview with Carol Cone, and company documents she provided.

Round-table discussion moderated by James Rippe: published as "Walking for fitness: A round table," *The Physician and Sports Medicine,* volume 14, 1986, pp. 145–59.

Walking-shoe statistics: John Shabe, "Taking fitness in stride," *Syracuse Post Standard,* June 26, 1987, p. D4.

Walking book: Robert Sweetgall, James Rippe, and Frank Katch, *Fitness Walking,* New York, Perigee, 1985; quotes appear on pp. 74–75.

Ladies' Home Journal **article:** Maureen Lynch, "The beauty of running," April 1978, p. 114.

George Sheehan quotes: *Dr. Sheehan on Running,* Mountain View, California, World Publications, 1975, pp. 14, 35.

Statistics on and reasons for giving up exercise programs: "Motivating people to keep exercising," *Optimal Health,* May/June 1987, pp. 32–34; Sabol, *The Body of America,* cited earlier, p. 134.

Exercise claims from the 1800s: Green, *Fit for America,* cited earlier, chapters 4 and 8; Donald J. Mrozek, *Sport and the American Mentality,* Knoxville, University of Tennessee Press, 1983, chapter 1.

Dr. Kenneth Cooper's claims: *The Aerobics Program for Total Well Being,* New York, M. Evans, 1982; Mildred Cooper and Kenneth Cooper, *Aerobics for Women,* New York, M. Evans, 1972.

Percent exercising: Joel Gurin and T. George Harris, "Taking charge," *American Health,* March 1987, pp. 53–57; Trish Hall, "Self-denial fades as Americans return to the sweet life," *The New York Times,* March 11, 1987, pp. C1, C8; *University of California Wellness Letter,* July 1987, p. 2.

Fred Stutman quote: *Walk Don't Die,* Philadelphia, Medical Manor Books, 1986; quote appears on p. 67.

Pittsburgh epidemiologists: Ronald E. LaPorte, Stephen Dearwater, et al., "Cardiovascular fitness: Is it really necessary?" *The Physician and Sports Medicine*, volume 13, 1985, pp. 145–50; quote appears on p. 148.

Kenneth Cooper's "While we're asleep . . .": Cooper and Cooper, *Aerobics for Women*, cited earlier, p. 13.

How much to exercise: Tom Shealey, "How much exercise do you really need?" *Prevention*, November 1985, pp. 59–63; *University of California Wellness Letter*, July 1987, p. 3; Margaret Pierpont, "The new 'soft' exercise," *Self*, January 1987, pp. 72–76. The U.S. Public Health Service takes a middle-of-the-road and somewhat open-ended approach. They define "appropriate physical activity" as "exercise which involves large muscle groups in dynamic movement for periods of 20 minutes or longer, 3 or more days per week, and which is performed at an intensity of 60 percent or greater of an individual's cardiorespiratory capacity" (Carl J. Caspersen, et al., "Status of the 1990 physical fitness and exercise objectives," cited earlier.)

Authorities quoted on amount of exercise: Arthur Leon's remarks appear in *Discover*, October 1984, p. 22; George Sheehan is quoted in Steve Johnson, "Running improves physical, mental well-being, studies show," Chicago Tribune news service, January 11, 1987; Kenneth Cooper is quoted in "New rules of exercise," *U.S. News and World Report*, August 11, 1986, pp. 52–56, and in Patricia Raber, "Aerobic exercise in perspective," *Rx Being Well*, November/December 1986, pp. 35–48; Henry A. Solomon quote is from his *The Exercise Myth*, New York, Harcourt Brace Jovanovich, 1984, pp. 62–63.

Studies supporting moderation: Dan Streja and David Mymin, "Moderate exercise and high-density lipoprotein cholesterol: Observations during a cardiac rehabilitation program," *Journal of the American Medical Association*, volume 242, November 16, 1979, pp. 2190–92; K. Magnus, A. Matroos, and J. Strackee, "Walking, cycling or gardening, with or without seasonal interruption, in relation to acute coronary events," *American Journal of Epidemiology*, volume 110, 1979, pp. 724–33; LaPorte, Dearwater, et al., "Cardiovascular fitness," cited earlier; quote appears on p. 147.

Marathon hypothesis quote: Thomas Bassler and Jack Scaff, "Letter to the Editor," *The New England Journal of Medicine*, volume 292, 1975, p. 1302.

Background factors protecting more active people from heart disease, and self-selection problems: In addition to Solomon's argument, see Ronald E. LaPorte, Lucile L. Adams, et al., "The spectrum of physical activity, cardiovascular disease and health: An epidemiological perspective," *American Journal of Epidemiology*, volume 120, 1984, pp. 507–17.

Paffenbarger study: Ralph Paffenbarger, R. T. Hyde, et al., "Physical activity, all-cause mortality, and longevity of college alumni," *The New England Journal of Medicine*, volume 314, 1986, pp. 605–13; and "A natural history of athleticism and cardiovascular health," *Journal of the American Medical Association*, volume 252, 1984, pp. 491–95.

Information on Paffenbarger as a long-distance runner, and his concern about the stairs question: *Jim Fixx's Second Book of Running*, New York, Random House, 1978, pp. 26–27.

Income statistics on Harvard alumni: *The Stanford Magazine*, Spring 1986.

Correlation and causation: Stephen Cole, *The Sociological Method*, Chicago, Rand McNally, 1976; Darrell Huff, *How to Lie with Statistics*, New York, Norton, 1954, chapter 8 (the rum example appears on p. 90); and see also Barry Glassner, *Essential Interactionism*, London, Routledge & Kegan Paul, 1980, chapter 1.

Marshall H. Becker quotes: "The tyranny of health promotion," cited earlier.

On exercise as weight-increasing: Schwartz, *Never Satisfied*, cited earlier, pp. 98–107.

Conflicting findings regarding exercise, and anxiety and depression: C. Barr Taylor, James F. Sallis, and Richard Needle, "The relation of physical activity and exercise to mental health," *Public Health Reports*, volume 100, 1985, pp. 195–202; Melvin J. Stern and Patricia Cleary, "The national heart exercise and heart disease project: Long-term psychosocial outcome," *Archives of Internal Medicine*, volume 142, 1982, pp. 1093–97; Bonnie Berger and David Owen, "Mood alteration with swimming," *Psychosomatic Medicine*, volume 45, 1983, pp. 425–32; Ferris N. Pitts, "Biochemical factors in anxiety neurosis," *Behavioral Science*, volume 16, 1971, pp. 82–91; Carlyle H. Folkins, "Effects of physical training on mood," *Journal of Clinical Psychology*, volume 32, 1976, pp. 385–88; K. Hardman, "A dual approach to the study of personality and performance in sport," in Harold T. Whiting, K. Hardman, and Derek Hendry (eds.), *Personality and Performance in Physical Education and Sport*, Lafayette, Indiana, Balt Publishers, 1973, pp. 77–122.

Satisfaction-with-life poll: Gurin and Harris, "Taking charge," cited earlier, p. 56.

John R. Hughes study and quote: "Psychological effects of habitual aerobic exercise," *Preventive Medicine*, volume 13, 1984, pp. 66–78.

Other review of studies: Carlyle H. Folkins and Wesley E. Sime, "Physical fitness training and mental health," *The American Psychologist*, volume 36, 1981, pp. 373–89.

Posture, smiles, and gait as mood improvers: Sara E. Snodgrass, "The effects of walking behavior on mood," paper presented at the meetings of the American Psychological Association, 1986.

Athletes coping with stress: David Sinyor, Sandra Schwartz, et al., "Aerobic fitness level and reactivity to psychosocial stress: Physiological, biochemical and subjective measures," *Psychosomatic Medicine*, volume 45, 1983, pp. 205–16; David Sinyor, Morrie Golden, et al., "Experimental manipulation of aerobic fitness and the response to psychosocial stress," *Psychosomatic Medicine*, volume 48, 1986, pp. 324–37.

Characteristics of exercisers: Charles Lupton, Nancy Ostrove, and Robert Bozzo, "Participation in leisure-time physical activity," *Journal of Physical Education, Recreation, and Dance*, volume 55, 1984, pp. 19–23; Burton Brunner, "Personality and motivating factors influencing adult participation in vigorous physical activity," *Research Quarterly*, volume 40, 1969, pp. 464–69; John Young and A. Ismail, "Ability of biochemical and personality variables in discriminating between high and low physical fitness levels," *Journal of Psychosomatic Research*, volume 22, 1979, pp. 193–99; Emile Farge et al., "Runners and Mediators," *Journal of Personality Assessment*, volume 43, 1979, pp. 501–3; Klaus Heinemann, "Unemployment, personality, and involvement in sport," *Sociology of Sport Journal*, volume 2, 1985, pp. 157–63; B. Diane Hayes, William Cockerham, and Gunther Luschen, "Exercise and well-being," paper presented at the annual meetings of the American Sociological Association, 1986.

Placebo effects, sociation, and other nonphysiological factors producing effects from exercise programs: Larry A. Tucker, "Obesity, exercise, somatotype, and psychological well-being," *Journal of Human Movement Studies*, volume 9, 1983, pp. 125–33; Robert Sonstroem, "Exercise and self-esteem," *Quest*, volume 33, 1982, pp. 124–39; Elaine Heiby et al., "A cognitive-behavioral model of adherence to health-related exercise," paper presented at the meetings of the American Psychological Association, 1985; George M. Andrews, Neil B. Oldridge, et al., "Reasons for dropout from exercise programs in postcoronary patients," *Medicine and Science in Sports and Exercise*, volume 13, 1981, pp. 164–68; Richard A. Heaps, "Relating physical and psychological fitness," *Journal of Sports Medicine and Physical Fitness*, volume 18, 1978, pp. 399–408; Folkins and Sime, "Physical fitness training and mental health," cited earlier; Berger and Owen, "Mood alteration with swimming," cited earlier; Solomon, *The Exercise Myth*, cited earlier, chapter 6.

Canadian Type-A study: Ethel Roskies et al., "The Montreal Type A intervention project," *Health Psychology*, volume 5, 1986, pp. 45–69. Further cautions about the power of exercise to relieve stress appear in Sinyor, Golden, et al., "Experimental manipulation of aerobic fitness," cited earlier; and in Thomas G. Plante and Dennis Karpowitz, "The influence of aerobic exercise on physiological stress responsivity," unpublished paper.

TWELVE: THE SECRET AGENDA

Psychology Today **survey:** Cash et al., "The great American shape-up," cited earlier.

On contemporary body improvement efforts as moral endeavors: Robert Crawford, "A cultural account of 'health,' " in John B. McKinlay (ed.), *Issues in the Political Economy of Health Care*, London, Tavistock, 1984; Schwartz, *Never Satisfied*, cited earlier; Turner, *The Body and Society*, cited earlier; John O'Neill, *Five Bodies*, Ithaca, New York, Cornell University Press, 1985; Susan Sontag, *Illness as Metaphor*, New York, Farrar, Straus & Giroux, 1978; Millman, *Such a Pretty Face*, cited earlier; Howard Stein, "Neo-Darwinism and survival through fitness in Reagan's America," *The Journal of Psychohistory*, volume 10, 1982, pp. 63–87; and " 'Health' and 'wellness' as euphemism: The cultural context of insidious draconian health policy," *Continuing Education for the Family Physician*, volume 16, 1982, pp. 33–44.

Emile Durkheim on morality: *Moral Education*, New York, Free Press, 1961; see especially pp. 120–21.

The diabetes book: Robert Cantu, *Diabetes and Exercise*, New York, E. P. Dutton, 1982.

Analysis and quoted material regarding diabetes, obesity, and diet: Schwartz, *Never Satisfied*, cited earlier, p. 173; Lena E. Goodman and Madeleine J. Goodman, "Prevention—How misuse of a concept undercuts its worth," *Hastings Center Report*, volume 16, 1986, pp. 24–25.

1986 Gallup poll: Gurin and Harris, "Taking charge," cited earlier, p. 54.

Class, race, and health-related behaviors: Dennis Gilbert and Joseph Kahl, *The American Class Structure*, Homewood, Illinois, Dorsey, 1982; Caspersen, et al., "Status of the 1990 physical fitness and exercise objectives," cited earlier; D. W. Edington, "Health behaviors tied to education level," *Optimal Health*, May/June 1986, p. 60; B. Diane Haynes et al., "Exercise and well-being," and Sanford M. Dornbusch, "Norms for thinness among adolescent females," both papers presented at the annual meetings of the American Sociological Association, 1986. See also a survey conducted in 1986 by Clark, Martire and Bartolomeo, Inc., for the American Cancer Society, "Diet, Nutrition, and Cancer," which found that 73 percent of blacks, compared to 51 percent of whites, believe that whether or not one gets cancer is determined by "God's will." This survey also found that blacks and Hispanics were less likely to have increased their consumption of fiber or decreased their intake of red meat. (The results of the survey appear in a report from the research organization.)

Health professionals' messages about eating habits: Marilyn G. Stephenson et al., "1985 NHIS findings: Nutrition knowledge and baseline data for the weight-loss objectives," *Public Health Reports*, volume 102, number 1, pp. 61–67.

Robert Crawford's study and quotes: "A cultural account," cited earlier.

Public Health Service figures: Caspersen et al., "Status of the 1990 physical fitness and exercise objectives," cited earlier; Stephenson et al., "1985 NHIS findings," cited earlier.

John F. Kennedy vigor quote: from *Sports Illustrated*, reproduced in Patricia A. Eisenman and C. Robert Barnett, "Physical fitness in the 1950's and 1970's: Why did one fail and the other boom?" *Quest*, volume 31, 1979, pp. 114–22. Also on the Kennedy era, see Donald J. Mrozek, "The cult and ritual of toughness in Cold War America," in Ray B. Browne, *Rituals and Ceremonies in Popular Culture*, Bowling Green, Ohio, Bowling Green University Popular Press, 1980, pp. 178–91.

Catherine Gallagher quote: "The body versus the social body in the works of Thomas Malthus and Henry Mayhew," *Representations*, issue 14, 1986, pp. 83–106.

Muriel R. Gillick's comments: "Health promotion, jogging, and the pursuit of the moral life," *Journal of Health Politics, Policy and Law*, volume 9, 1984, pp. 369–87.

Jane Fonda quote: *Jane Fonda's Workout Book*, New York, Simon and Schuster, 1981, p. 47.

Rita Freedman quote: *Beauty Bound*, cited earlier, p. 166. See also Wendy Chapkis, *Beauty Secrets*, Boston, South End Press, 1986, pp. 12–13.

Ivan Illich quotes: *Medical Nemesis*, New York, Pantheon, 1975, p. 1; "Body History," *The Lancet*, December 6, 1986, pp. 1325, 1327.

ACKNOWLEDGMENTS

My thanks go first to the interviewees for sharing their time, experiences, and feelings with me.

Three people have supported this project enthusiastically from the start. Betsy Amster, my wife, not only expertly critiqued and helped revise several drafts of each chapter but endured my highs and lows along the way. Geri Thoma, my agent, was always available for encouragement and good conversation. Faith Sale, my editor, made the book stronger—and thinner—with her careful editing.

For helpful discussions, I want to thank Howard Becker, Charlotte Kahn, Hal Mizruchi, Jonathan Moreno, Barry Schwartz, and Eviatar Zerubavel. Thanks also to those who assisted with the research: Cheryl Carpenter, Victoria Guevara, Rebecca Riehm, and librarians at Syracuse University and SUNY Upstate Medical Center libraries.

Index

Benson, Harry, 134
Biceps, 118, 125
Biking, 96, 201, 227
Binge eating. *See* Bulimia
Blacks, 16, 65, 277
  cosmetic surgery and, 193
  positive self-image of mothers
    among, 76–81
  stereotyping of, 21, 144–51
  See also Race
Blepharoplasty, 188
Body image. *See* Ideal body images;
  Self-image
Body remakers. *See* Beauty make-
  overs; Electrology; Health
  clubs; Plastic surgery; Weight-
  loss programs
Bodybuilding, 123, 256, 267
  See also Muscles
Bordo, Susan, 89
Boxing, 124, 141–43
Breasts, 78, 95–96
  cancer of, 196
  ideal images of, 12, 45
  surgery on, 19, 59–61, 188, 190–
    91, 197
Brody, Jane, 156
Bulimia, 56, 58, 89–91, 98, 100, 101,
  265
Business. *See* Corporations
Buttocks, 79
  ideal images of, 12, 18, 45
  plastic surgery on, 111, 197
  skin grafts from, 18–19

Cancer, 52, 93, 196, 240, 277
Career. *See* Work/career
Catholics, 116
Causation vs. correlation, 240–41, 275
Chapkis, Wendy, 107
Cher, 12, 215
Chernin, Kim, 89–90
Chicanos, 125
Children
  approaches to rearing of, 94
  attractiveness in, 33–34, 82–83,
    262, 264–65
  beauty ads and, 46, 47, 262–63
  choosing to have, 99–100

dieting among, 47, 263
  wife's career vs., 164–66, 169
  See also Pregnancy
Chodorow, Nancy, 67, 72
Cholesterol, 164, 241, 274
Class. *See* Social class
Cleanliness. *See* Self-purification
Clothes, 24, 25, 27–30, 32, 38, 106
  men's concern with, 20, 137–40,
    142, 143, 150
Coleman, James, 113
College, 53–55, 58–59, 69, 78–79, 83–
    84, 105, 110, 127, 142, 148
  attitudes toward athleticism at, 118,
    120
  eating disorders in, 46, 58
  weight and admission to, 52
Collins, Joan, 134, 136
"Commodification," 260
Commodities. *See* Products
Cone, Carol, 227–32, 234
Connecticut, health clubs in, 206
Conrad, Peter, 223–24
Constipation, 233–34
Consumption cues, 29
  See also Products
Control/self-control, 13, 89, 29–94,
    96, 97
  as class-specific concept, 251
  daughters of alcoholics and, 101–6
  and marriage, 98–100
  in prison, 123–28
  social forces and, 106–7
  women's lack of social power and,
    106–11
Cooper, Kenneth, 233–37
Corporations, 13, 201
  health promotion by, 13, 219, 221–
    24, 271–72
  and public health, 224
  women in, 108–11
Correlation vs. causation, 240–41, 275
Cosmetic surgery. *See* Plastic surgery
Cosmetics, 13, 14, 30, 32, 38, 122,
    177
  See also Makeup

Dancing, 84–87, 104, 227, 235
  body type and, 44